Adhesives and Consolidants in Painting Conservation

Adhesives and Consolidants in Painting Conservation

Edited by
Angelina Barros D'Sa, Lizzie Bone,
Rhiannon Clarricoates and Alexandra Gent

Archetype Publications

in association with
the Icon Paintings Group

First published 2012 by Archetype Publications Ltd in association with the Icon Paintings Group

Archetype Publications Ltd
c/o International Academic Projects
One Birdcage Walk
London SW1H 9JJ
www.archetype.co.uk

ISBN: 978-1-904982-88-3

British Library Cataloguing in Publication Data
A catalogue record for this book is available from the British Library.

Printed on acid-free paper

Designed by Marcus Nichols at PDQ Digital Media Solutions Ltd.
Typeset by PDQ Digital Media Solutions Ltd, Bungay
Printed and bound in Great Britain by Latimer Trend & Co. Ltd., Plymouth

Contents

Foreword

In 2010 the Committee of the Icon Paintings Group decided to organise a conference on adhesives and consolidants in painting conservation. The papers selected for the conference included contributions from eight different countries.

Adhesives and consolidants are undoubtedly of great interest to conservators across all specialisms, and the conference *The Sticking Point: Adhesives and Consolidants in Painting Conservation*, held at the National Portrait Gallery in London in May 2011, was understandably fully subscribed.

Whilst such materials are a fundamental part of conservation, adhesives and consolidants by their nature are intrusive and present many concerns for the conservator. All those involved in conservation have seen the problems that can be encountered with adhesives and consolidants used in the past – brittleness, yellowing and discoloration, and the difficulty in completely removing an old adhesive. There is now so much emphasis on preventive conservation and minimal intervention that the application of an adhesive or consolidant is not carried out without first, considering the necessity, and second the choice of an effective, safe and stable substance.

Discussions, symposiums and conferences allow conservators to share problems encountered with the practical application of materials or to be shown new uses for recognised materials. New materials and the results of their application always arouse interest. We need to know the limitations of materials and the possible implications of their use so we can make judgements and informed choices when tailoring treatments.

Needless to say the conference created a great deal of interest and this publication of the papers is the result of its success.

Francis Downing
Chair of the Icon Paintings Group Committee

Editor's note

In addition to the ten papers published in this volume, two further papers were presented at the conference: 'Synthetic binding media a necessary choice: the background to the development of the Lascaux Medium for Consolidation' by Hans Peter Hedlund and 'A working group on consolidation: investigation, research, decision making processes in treating a sensitive painting by Mario Sironi: results of a project 2004–8' by Matteo Rossi Doria, Ezio Buzzegoli and Diane Kunzelman. The research underpinning these papers has been published elsewhere (Hedlund and Johansson 2005; Buzzegoli *et al.* 2008).

References

Hedlund, H.-P. and Johansson, M. (2005) 'Prototypes of Lascaux's Medium for Consolidation: development of a new custom-made polymer dispersion for use in conservation', *Restauro* 6: 432–9.

Buzzegoli, E., Cavalca, N., Falucci, C., Kunzelman, D., Landi, L., Minotti, D., Marzullo, S., Mastropasqua, P., Michalski, S., Picollo, M., Rossi Doria, M., Tellini, S. and Tosini, P. (2008) 'Open Studio e intervento di restauro su un'opera di Sironi', in *L'Attenzione alle superfici pittoriche. Materiali e metodi per il consolidamento e metodi scientifici per valutarne l'efficacia – 2: atti del congresso, Milano, 21–22 novembre 2008, a cura di CESMAR7 e Diane Kunzelman. Quarto congresso internazionale, Colore e conservazione materiali e metodi nel restauro delle opere policrome mobili.* Il Prato: Saonara.

Acknowledgements

With thanks to all of the contributors, peer reviewers and editors whose hard work has made this publication possible.
We very much hope that you enjoy it, and that it is a useful resource for you.

The ICON Paintings Committee

Setting the scene: considerations and decision making for planning and undertaking treatments

Velson Horie

Introduction

A variety of materials has been used to preserve objects for millennia (Figure 1). Initially, conservators used whatever was to hand, but increasingly they searched for better materials. In the past few decades, materials have started to be made specifically for conservation use, but to be able to make a rational judgement about which material to use, the important criteria must be determined in order to inform the judgement. Figure 2 illustrates a range of criteria that might be considered. However for each task, indeed each object, the criteria relevant to its conservation have to be agreed. Once the criteria have been listed, the decision has to be made as to the importance of each one, how it is to be assessed and what is an acceptable value for that criterion.

What is an 'adhesive'?

Unfortunately, the vocabulary we have at our disposal is limited. We adopt words and meanings from common speech and then try to apply more specific conservation meanings to them. One example is 'adhesive', a term that is used to describe two different things: the first is the liquid applied to objects and the second is the solid matter that holds substrates together. As the same word is used to describe these two different materials, we tend not to distinguish the change of properties that occurs during the transformation from one state to another. It is therefore less ambiguous to use explicitly different terms and definitions for the two materials (and comparable definitions for consolidants and coatings) (Horie 2010):

- *Liquid adhesive*: a liquid adhesive, when applied to a substrate, sets to form an adhesive.
- *Adhesive*: an adhesive is a substance that holds two surfaces together by interfacial forces.

Each discipline has its own set of criteria and conservation is no different. An outline set of criteria for conservation might be:

- The liquid adhesive must flow over the surface.
- The liquid adhesive must set to become immobile.
- The adhesive must adjust to stresses and strains.
- The adhesive must be removable.

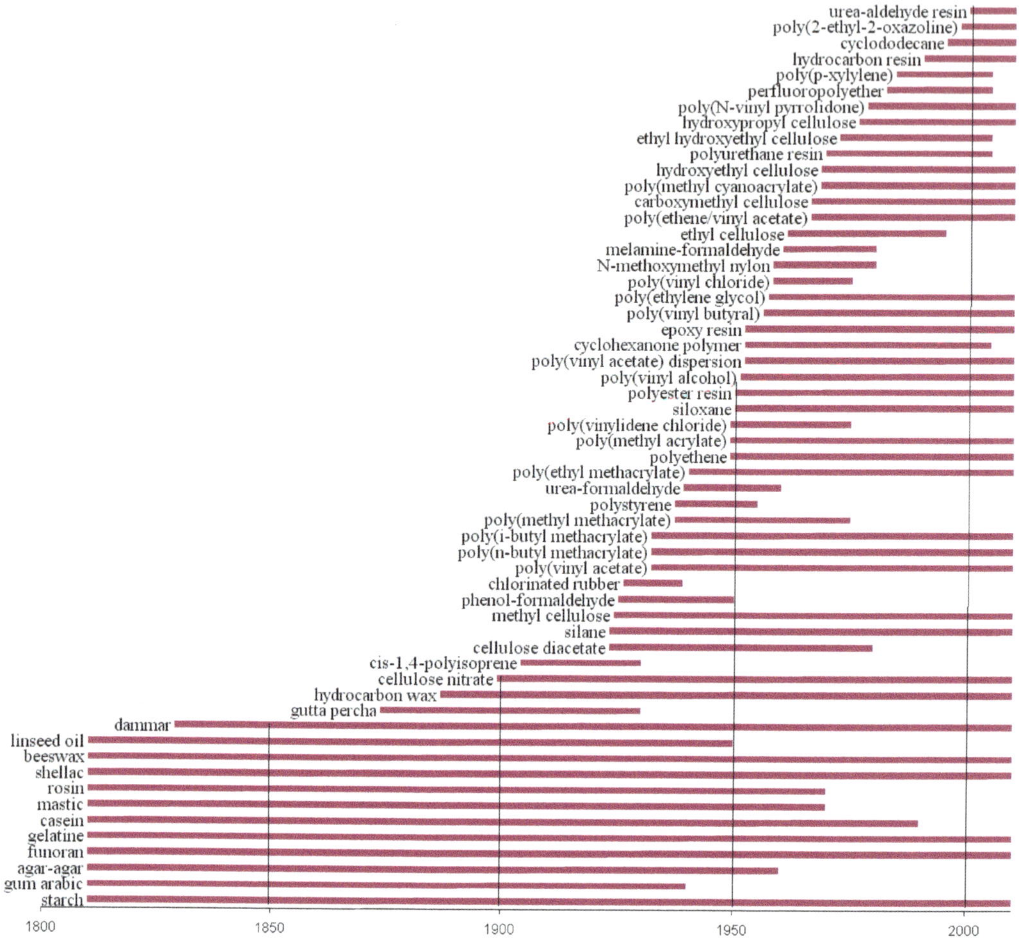

Figure 1 Materials have a long and varied history in conservation. This chart summarises published information about the period of introduction, use and abandonment of such materials. (Data from Horie 2010.)

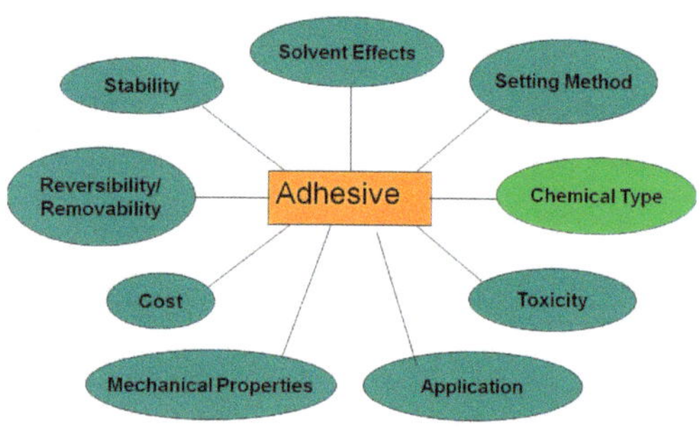

Figure 2 Criteria for some of the properties that should be considered when choosing an adhesive. A comparable list for the object should be drawn up, including ethical, financial and other factors.

Adhesion

The first stage in creating a bond is ensuring that the liquid adhesive flows over the surface. This process is influenced by a number of properties of both the liquid adhesive and the substrate (the object). If the liquid is not attracted to the substrate surface, as is the case with water on wax, it will ball up into droplets that sit on the surface without wetting it (Figure 3, top). If the liquid is attracted to the surface, for example acetone on varnish, the drop will spread spontaneously (Figure 3, bottom).

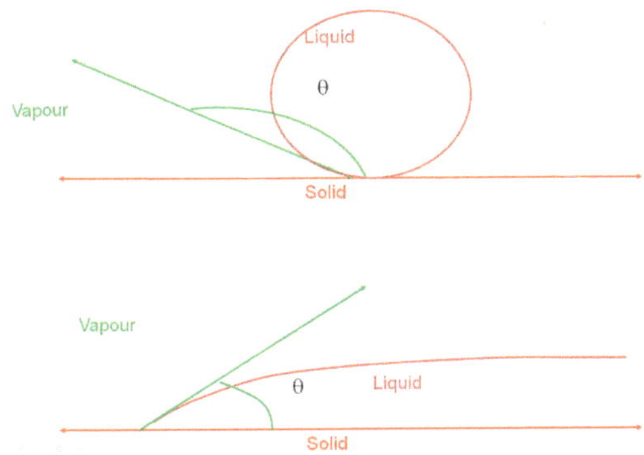

Figure 3 There is a balance of forces at the point of interaction between the liquid adhesive and the substrate. This balance can be summarised in the angle of interaction, which can be observed in practice. This allows the conservator to do a quick check that the adhesive will indeed spread and wet the surface.

Technically the attraction can be measured as the angle of contact (θ): the smaller the value, the better the wetting ability. Liquids with high surface tension, such as water or mercury, have high angles of contact, whereas liquids with low surface tension, such as petroleum or silicones, have low contact angles. As conservators, we can adjust the properties of both the liquid adhesive, for example by changing the solvent, and the substrate using a variety of treatments. Figure 4 shows how 'cleaning' a wood surface by abrasion (sanding) improves wetting by water.

These are static images of wetting, but in practice wetting is a process. The liquid has to flow, and that is affected by the viscosity of the liquid and the physical barriers on the substrate it has to overcome. A high viscosity liquid may eventually – in a matter of days or years – flow out over a surface, whereas a low viscosity liquid will flow very quickly. The flow of the liquid is also affected by changes in the liquid adhesive over time. For instance, a solvent carrier might progressively evaporate or a chemical reaction such as cross-linking may be taking place in the liquid adhesive, thus increasing the viscosity and changing the surface tension of the liquid. The substrate may be smooth or rough, or it might be porous. These irregularities affect both the ability and the speed at which a liquid can flow over and wet the surface.

Setting

The liquid adhesive then has to set into a solid adhesive, a transformation that involves considerable changes in the applied material. For instance, a 2% polymer solution will lose

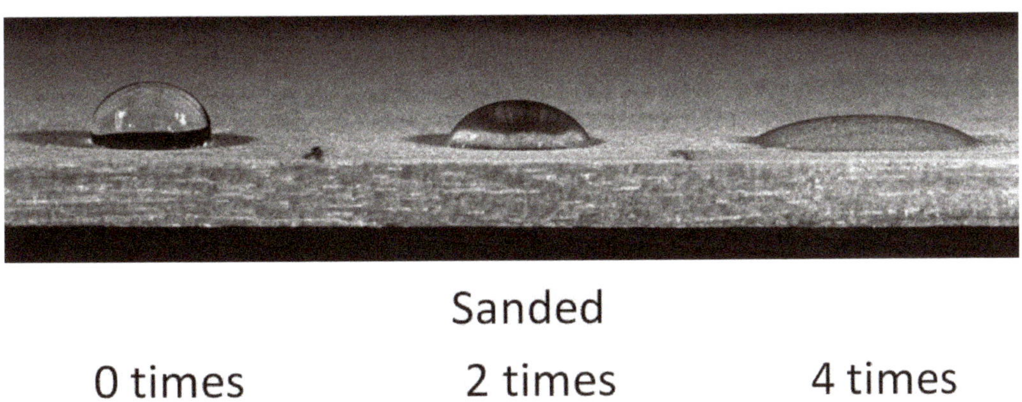

Figure 4 Wetting of a wood surface by a drop of water after various surface treatments. (Reproduced by kind permission from C.R. Frihart, 'Wood adhesion and adhesives', in R.M. Rowell (ed.), *Handbook of Wood Chemistry and Wood Composites*, 2005, CRC Press, pp. 216–78.)

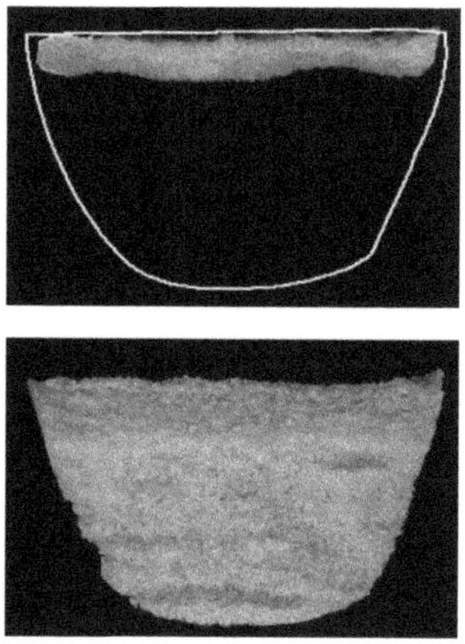

Figure 5 An ice cube mould was filled with loose sand mixed with a consolidant solution. This was allowed to set by evaporation. The mould was then dumped out and the loose sand brushed away – the sand that remained had been consolidated. It can be seen that the consolidation is not even, and the polymer's final distribution is highly dependent on the initial concentration of Paraloid B-72 in acetone. This behaviour is commonly observed for reasons as yet unknown. (Figures kindly provided from G.W. Brown, 'Ineffective consolidation: poor penetration or peripatetic polymer?', *Journal of Vertebrate Paleontology* (2002) 22(3a): 39A.)

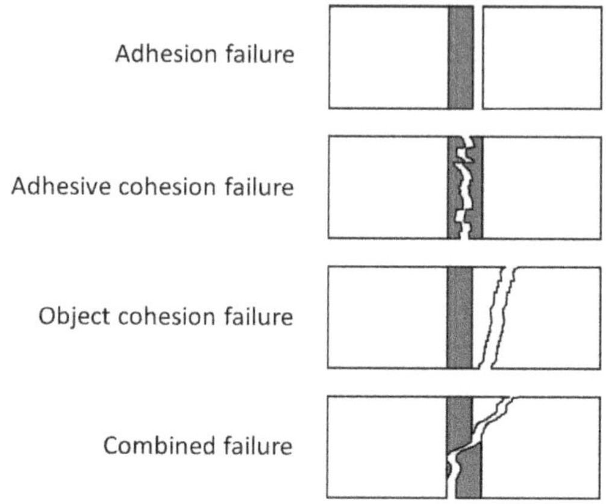

Figure 6 An adhesive bond can fail in various modes. From the conservation point of view adhesion failure is preferable but in industrial applications this is the worst possible mode. Making use of industrial research can therefore be misleading, since the fundamental aims of the research are so different. (From an original by Jane Down, Canadian Conservation Institute.)

98% of its volume on drying to a film. Figure 5 shows how a polymer can redistribute in a substrate depending on the original concentration of the polymer solution.

Effect of the added material

The purpose of applying an adhesive or consolidant is often to change the physical properties, usually to increase strength. In industry, the aim is for an adhesive stronger than the substrate, as illustrated by 'object cohesion failure' in Figure 6, but the converse applies in conservation, where the most desirable outcome at the time when failure eventually occurs is 'cohesive failure of the adhesive'.

Considerable research has been carried out into how a solid adhesive interacts with a substrate. It is important, for example, when attaching wings to aircraft. Adding an adhesive into a system causes considerable differences to the stresses and strains in an object (Figure 7). It is possible to model stresses and strains mathematically using known properties of the adhesive and adherend (i.e. the object) using the technique of finite element analysis. This is used in an industrial situation such as aircraft design and is starting to be applied in conservation.

When stresses are applied to a joint, even greater stresses can be generated at the adhesive interface than those originally applied, thus increasing potential for damage. Clever design of the joint is critical to reducing these maximum forces to below the (unknown) value that the object can withstand.

Ageing

The adhesive bond can change in many ways. The solid adhesive might yellow without affecting adhesive strength, it might become brittle, or it might peel off the substrate. The substrate might change as well, and differently.

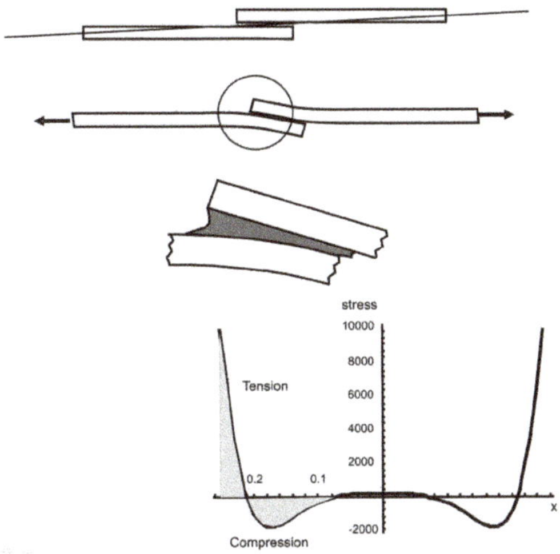

Figure 7 When an object is stressed, its response may not conform to expectations. This diagram shows one model of two metal strips joined by a strong, flexible adhesive. The resultant strain causes distortion in peculiar places, and concentrates forces at specific points, not evenly distributed. The adhesive tends to peel away from the metal, which is why adhesive failure is so damaging to the joint. Similar predictions could be made for conservation joins, which have rarely been (but should be) explored in this detail. (Drawing kindly provided by David Dillard, Virginia Polytechnic Institute and State University.)

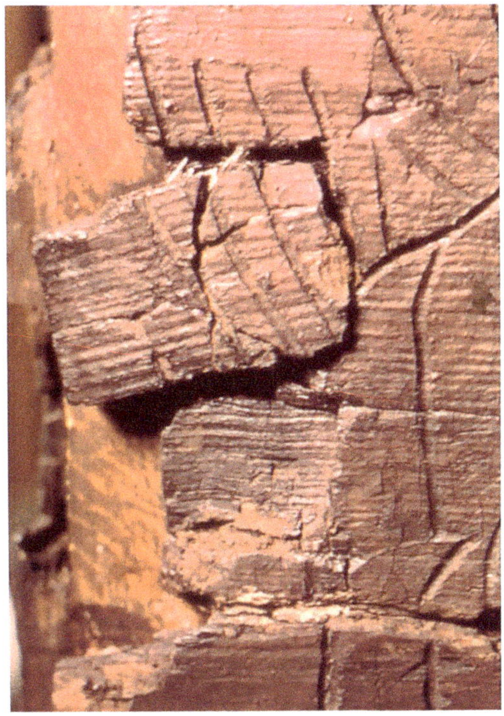

Figure 8 The wood fragments of St Cuthbert's coffin were stuck together in the 1940s using poly(vinyl acetate) adhesive. Over the subsequent 30 years, this fragment slumped due to the slow flow of the polymer. Its glass transition temperature was too low for this use, so it behaved as a high viscosity liquid. (From J.M. Cronyn and C.V. Horie, *St. Cuthbert's Coffin: The History, Technology and Conservation*, 1985, Dean and Chapter, Durham Cathedral.)

One example of long-term change, creep, is a property of many thermoplastics that behaves as a very high viscosity liquid, as illustrated in Figure 8. One of the criteria for choosing conservation materials are the long-term changes, both physical and chemical, including those changes that may result as a consequence of the object's use.

The unsticking point

Available literature and the received wisdom of conservators describe treatments that were once applied with enthusiasm, and which we now have to remove, often with considerable difficulty. Part of the planning process for a conservation treatment must be preparing for the reversal phase. Unfortunately and frequently, reversal requires very different methods from those used for application. For instance, polymers applied from dispersions diluted with water (selected perhaps because the adherend was sensitive to solvent action), require considerable amounts of organic solvents for swelling and washing out the polymer. In the process, the polymer will expand, causing fractures in the object at micro and macro levels (Figure 9).

Recommendations

Any treatment carried out now will have long-term consequences. When designing a treatment for an object, 'thinking through'

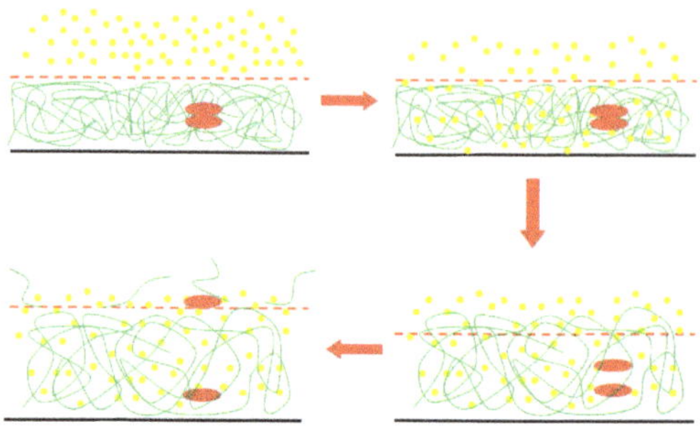

Figure 9 Dissolving a polymer film requires that a polymer molecule must be completely freed from its matrix in order to flow away with the solvent. This in turn means that the polymer matrix swells to allow the solvent to penetrate and surround the molecules. Any object encapsulated in the polymer film will be disrupted and liable to be washed away in the solvent. This is the reason why pigment particles are removed when a consolidant is removed, for instance when using a poultice. (Redrawn from K. Ueberreiter, 'The solution process', in J. Crank and G.S. Park (eds), *Diffusion in Polymers*, 1968, Academic Press, pp. 219–57.)

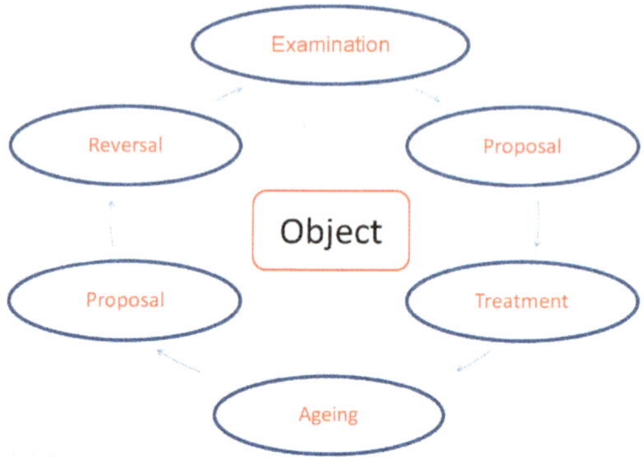

Figure 10 Objects undergo/suffer cycles of treatment over the centuries of their use and conservation. Conservators must explicitly plan for the re-conservation of their objects In the treatment plan, including where possible a planned timetable for this re-treatment. Only when details of this cycle have been researched and documented for action by future conservators should the initial treatment be undertaken.

each stage of the process (Figure 10) will help to avoid obvious flaws and major difficulties both for our successors and the objects:

- We can never assume that we are right – such assumptions have a long history of being wrong. Figure 1 illustrates the many natural and synthetic materials that are no longer used for treating objects, and there were good reasons for discontinuing their use.
- Assumptions should always be checked – many are simply 'bad habits' or misapplied knowledge of how materials behave in not quite relevant circumstances.
- Good records should be made of the reasons behind the decision to carry out a particular treatment – these should be kept for future referral.
- Methods and materials should be tested and revalidated regularly.

Bibliography and references

Anon. (1992) *Adhesives and Coatings*, vol. 3. London: Routledge.

Horie, C.V. (2010) *Materials for Conservation: Organic Consolidants, Adhesives and Coatings*. Oxford: Taylor and Francis.

Pocius, A.V. (2002) *Adhesion and Adhesives Technology: An Introduction*. Munich: Carl Hanser Verlag.

Author's address

Velson Horie (www.horie.co.uk / info@horie.co.uk)

A look at the structural requirements of consolidation adhesives for easel paintings

Marion F. Mecklenburg, Laura Fuster-Lopez and Silvia Ottolini

Introduction

Successful conservation treatments are a combination of a carefully planned action together with the accurate selection of materials and methods. Such an apparently obvious statement necessarily implies:

- Identifying the nature of the damage.
- Analysing what caused the damage.
- Understanding the failure mechanism that has taken place.
- Determining (and prioritising) the needs.
- Selecting the most appropriate methods and materials.

In this sequence chemical, physical, mechanical and biological processes are usually involved. In the case of paintings, in addition to the chemical reactions responsible for the ageing and degradation of artworks, the mechanical and dimensional properties of the different materials determine how paintings behave structurally, how the different materials of the composite structure interact with each other and how such properties change in time, both as the painting ages and as a function of environmental oscillations.

Polymers are used widely in many conservation treatments to strengthen, adhere and protect. Factors such as their composition, molecular weight, glass transition temperature (T_g), strongly determine their viscosity, film-forming properties and long-term chemical stability, which clearly influences their suitability for each specific treatment. This paper focuses on some structural issues of adhesives and consolidants in painting conservation as well as the cohesion and adhesion forces governing their performance. Also discussed are their mechanical properties, their interaction with environmental conditions and how their ability to penetrate determines their role in the context of the painting structure.

Consolidation treatments in easel paintings

Defining the concept

The term 'consolidation' may in one interpretation refer to any treatment consisting of the reinforcement and repair of a damaged support and in another to the re-adhesion of weak and flaking paint to an underlying substrate. Equally confusing is the fact that similar (if not the same) adhesives and solutions have been traditionally used, apparently indiscriminately,

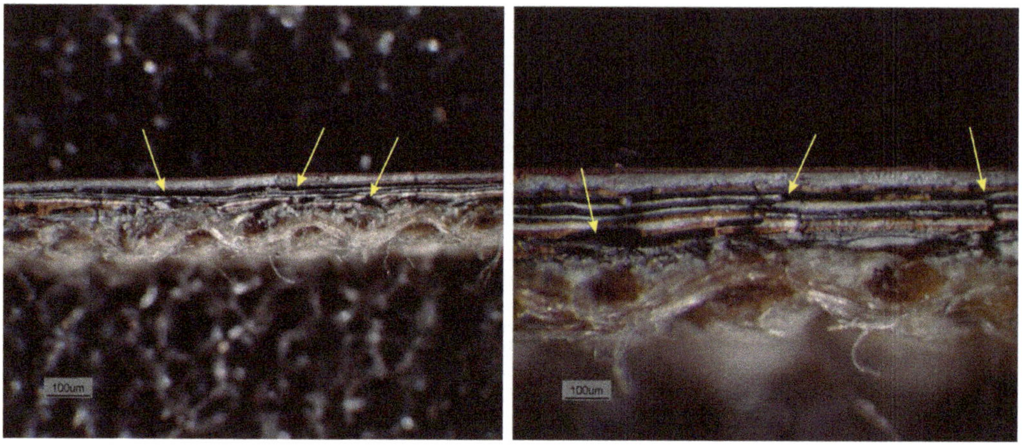

Figure 1 Cross-sections of a typical canvas painting showing evident delamination. The yellow arrows indicate the positions of the voids and, therefore, where the adhesive needs to penetrate. In general, adhesives must penetrate very small pore sizes. It is also clear that some difficulties will be faced to ensure that the adhesive reaches those voids as no cracks are observed on the surface. (Photos courtesy of Melvin J. Wachowiak.)

in line paintings to consolidate powdery paint layers, to impregnate weak canvases from the reverse or simply to adhere locally delaminated paint fragments. At this point, several questions arise:

- Are the strength requirements for adhesives used to consolidate wood the same as for those used to reattach flaking paint and for those used to impregnate a canvas?
- What are the properties required of adhesives in each case? More specifically, what are the differences between adhesives and consolidants in terms of adhesive and cohesive forces?
- Can, therefore, any adhesive be used as a consolidant?

Examining the behaviour of each layer of a painting as a function of the forces acting inside the composite structure helps to understand that the requirements in each case are not the same.

For canvas paintings, for example, lining and consolidation adhesives have different requirements since they address different needs. Adhesives for lining must present cohesive and adhesive properties whereas consolidating adhesives require adhesion and penetration properties. In the case of linings it is obvious that adhesion is needed to bond a new canvas to a painting. But also, since the main forces acting inside the structure of a traditional painting are usually concentrated in the size layer which acts as the primary support, it becomes clear that cohesive properties are equally desirable in lining adhesives in order to confer cohesion, strength and proper stiffness. Consolidating adhesives also require adhesive strength since paint fragments need to be bonded together. However, it must be noted that the failure of some consolidation treatments may be the result of poor adhesive penetration as well as weak adhesive properties.

Cracks, cleavage, cupping, blisters, flaking and delamination correspond to different failure mechanisms as a consequence of multiple factors, such as the properties of the materials, their ageing, as well as their interaction with environmental agents. Different failure mechanisms determine the location of the damage and how much void any adhesives have to fill (Figure 1). Success also depends on how well adhesives penetrate through the design layers to reach the point where they are really needed.

Defining the structural requirements of consolidating adhesives

To determine the structural requirements of adhesives, a consideration of the cohesive and adhesive properties of materials is required to find answers to questions such as:

- How much strength is required to consolidate powdery paint?
- Should adhesives be stiffer than the paint film?
- Do different solvents determine different mechanical properties for adhesives?
- Why are adhesives that show lack of strength in experimental testing considered to work well in practice?
- What is meant when describing an adhesive as 'working well'?
- Are adhesives reaching the optimum position to be effective or are current consolidation adhesives forming separate layers on top of which paint fragments remain unsecured?

Tests run during the course of this research were devised to determine the cohesive and adhesive behaviour of some of the adhesives currently used as consolidants. The penetration of various adhesives was also tested.

Cohesion is a measure of the mechanical properties of the adhesive itself and reflects the ability of the adhesive alone to develop stiffness and strength. The cohesive test is an excellent way to examine the drying time and the effects of residual solvents. It is also extremely useful in determining the effects of different solvents on the properties of the material. For this purpose, the standard stress–strain test in tension was used.

Adhesion is the material's ability to bond together different materials. Simple 'peel' tests were used in this series of tests. Various materials were bonded together and the resistance in separating the bonded substrates or 'adherents' was used as a measure of the adhesive strength.

Penetration of adhesives is an essential prerequisite for bonding. Both penetration and adhesion rely heavily on several factors that depend on:

- The substrate
  - The substrate's porosity and composition
  - The substrate's surface uniformity
  - The existence of previous coatings applied to the substrate, etc.
- The adhesive
  - Type of polymer
  - Molecular weight and size
  - Glass transition temperature
  - Solvent used for dilution
  - Concentration
  - Adhesive and cohesive strength
  - Dry down, or the loss of volume of the adhesive during drying
  - Drying time
  - Adhesive wetting of the substrate surface
  - Viscosity
  - Thixotropy
  - Application methodology, etc.
- Environmental conditions

This research was intended to determine the pore size that each adhesive is able to penetrate as a function of the solvent and the concentration used. For this purpose, various filter papers with different pore dimensions were used. Alone the filter papers were easily peeled apart by attaching tape to the outer surfaces. If however the papers were first dipped into selected adhesives applied with various solvents at different concentrations, the increased force required to 'split' the paper became a measure of their effectiveness in penetrating different pore sizes.

Some other observations could also be made regarding:

- The effects of solvents on film forming.
- How solvents determine the mechanical properties of a polymer.
- How solvents determine drying times.
- The consequences of residual solvents in the polymer in terms of the strength and stiffness of the resulting film.

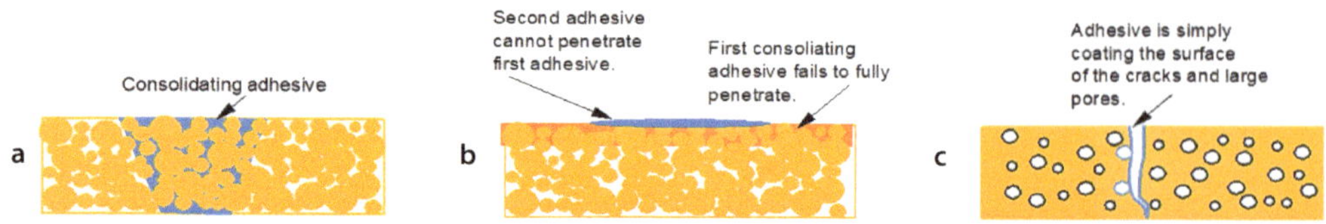

Figure 2 (a) This shows an adhesive that has fully penetrated a porous substrate; (b) this shows that the initial adhesive application can block any subsequent application; (c) this illustrates that the adhesive solution is able to penetrate the substrate but it simply coats the surfaces of the gaps and pores without bridging and bonding the material.

- The pore size that each adhesive can penetrate, depending on the resin, on the solvent and the concentration used.

Porosity, absorbency and penetration

Porosity is a measure of the number and size of pores/voids (empty spaces) in a material and it can also refer to how quickly a fluid can circulate through it: the lower the porosity, the slower the migration of the fluid through the pores. *Permeability* can be considered as a state where the small pores/voids are interconnected so that a fluid can pass from pore (or void) to pore. Both porosity and permeability determine the *absorbency* of a material. Absorbency is the ability of a material to retain a fluid. Generally the larger the particle size of the material, the more porous it is. Large particle-size substrates containing particles with a high surface area will absorb fairly well.

The aim of any consolidation treatment is to make the selected adhesive fully penetrate the paint layer and provide the lost adhesion. Most of the adhesives used in painting conservation set by evaporation of the solvent. Ideally the adhesive in solution or dispersion in water would penetrate the substrate only, the solvent or water would evaporate and the polymer (adhesive) would be left behind as seen in Figure 2a. It has to be taken into account, however, that canvas and the wood of panel paintings are extremely hygroscopic, porous and absorbent materials that tend to rapidly absorb solvent, thereby increasing the concentration and viscosity of the remaining adhesive, which then tends to block the surface (Figure 2b). It is interesting to note that even though this effect takes place quickly and can be interpreted as a fast-drying adhesive, research has shown that some solvent is retained in the substrate materials for a considerable amount of time.

Many consolidation treatments are intended to bond paint fragments together. The problems to consider range from powdery substrates to big paint flakes. In all cases however, adhesive has to connect the gaps and/or pores that exist in the material. Figure 2c shows an adhesive solution in a concentration that is too low. The solution penetrates the substrate and coats the pore walls without bridging the materials that need to be adhered. The loss of volume of the adhesive during drying (known as 'dry down') also contributes to this effect. Several applications of the adhesive might be necessary in order to provide an effective bond.

Experimental

Materials

In 2007 a survey was conducted in order to collect information on the most common adhesives used for consolidation purposes. The answers received are shown in Table 1 and the substrates selected for bonding tests are specified in Table 2.

Table 1 Adhesives selected for the experimental testing.

Adhesive	Solvent	Solution	Comments
Hide glue	Water	5%	Natural animal glue
Hide glue	Water	10%	
Sturgeon glue	Water	5%	Natural animal glue
Sturgeon glue	Water	10%	
JunFunori	Water	6%	Seaweed (*Gloiopeltis*)
Acril AC33	–	Undiluted (46%)	Acrylic dispersion based on ethylacrylate + methyl methacrylate
Acril AC33	Water	1:1	
Plextol B500	–	Undiluted (50%)	Dispersion: ethyl methacrylate and ethyl acrylate copolymer
Plextol B500	Water	1:1	
Plextol B500	Xylene	40:60 (w/w)	Dispersion: ethyl methacrylate and ethyl acrylate copolymer
Klucel G	Water	20%	Hydroxypropylcellulose
Klucel G	Water	8%	
Plexisol P550	–	Undiluted (40%)	Hydrocarbon solution of an acrylic polymer
Plexisol P550	Toluene	1:1	Hydrocarbon solution of an acrylic polymer
Aquazol 500	Water	10%	Polyoxazoline: thermoplastic polymers formed by poly(2-ethyl-2-oxazoline)
Aquazol 200	Water		Polyoxazoline: thermoplastic polymers formed by poly(2-ethyl-2-oxazoline)
Beva 371	Toluene	1:1	Ethylenevinylacetate, paraffin, ketone resin
Beva 371	–	Undiluted	Ethylenevinylacetate, paraffin, ketone resin
Paraloid B-67	Acetone		Isobutyl methacrylate polymer
Paraloid B-67	Methanol		Isobutyl methacrylate polymer
Paraloid B-67	Toluene		Isobutyl methacrylate polymer
Paraloid B-72	Different solvents		Methyl methacrylate copolymer
Paraloid B-72	Toluene	25% (w/w)	Methyl methacrylate copolymer
Paraloid B-72	Toluene	50% (w/w)	
Gelvatol	Water	10% (w/w)	Poly (vinyl alcohol)
Gelvatol	Water	20% (w/w)	Poly (vinyl alcohol)
Vinavil 59	–	Undiluted (42%)	Vinyl acetate
Vinavil 59	Water	1:1	Vinyl acetate
Vinavil NPC	–	Undiluted (51%)	Vinyl acetate
Vinavil NPC	Water	1:1	Vinyl acetate
Fluoline HY	–	Undiluted	Copolymer of vinylidene fluoride-esafluoropropene
Fluormet A	–	Undiluted	Fluoroelastomers and acrylic polymers
Elvacite 2044	Toluene	25% (w/w)	N-butyl methacrylate polymer
Elvacite 2044	Toluene	50% (w/w)	N-butyl methacrylate polymer
Elvacite 2046	Toluene	25% (w/w)	Iso-butyl / n-butyl methacrylate copolymer
Elvacite 2046	Toluene	50% (w/w)	Iso-butyl / n-butyl methacrylate copolymer
Akeogard T35	–	Undiluted	Aliphatic polyether-urethane
Akeogard T35	Water	1:1	Aliphatic polyether-urethane
Akeogard T40	–	Unidluted	Aliphatic polyether-urethane
Akeogard T40	Water	1:1	Aliphatic polyether-urethane

Methods and sampling

Tensile tests

The cohesion tests were simple tensile stress–strain tests of samples of pure adhesive cast in thin films on polyester sheets. The rate of loading was 30 seconds between test points and all the samples were equilibrated to the test environment for at least 24 hours prior to testing. The test environments are indicated on the graphs. The testing equipment was a purpose-built group of small tensile devices for measuring thin films in controlled environments. The relative humidity (RH) was controlled using trays of buffered silica gel and monitored using two-stage chilled mirror dew-point meters.

Samples for cohesive testing were prepared by either diluting the adhesive in the solvents listed or using the adhesive as supplied and

Table 2 Substrates selected for the adhesive and penetration testing (original magnification 40×). (Photos courtesy of Melvin J. Wachowiak.)

Substrate	Characteristics		Maximum pore dimension (in µm)	Thickness (in mm)	Adhesive free splitting force (N/cm)
	– Whatman paper is nearly pure cotton and has a known porosity. – It has a fairly soft surface texture. – It is used as a filter paper and stops particles 11 µm and greater. – The biggest difference between the Whatman #1 paper and the Kraft paper was the fibre size and surface finish.	Whatman #1	11	0.178	0.43
		Whatman #4	20–25	0.216	0.42
		Whatman #5	2.5	0.216	0.58
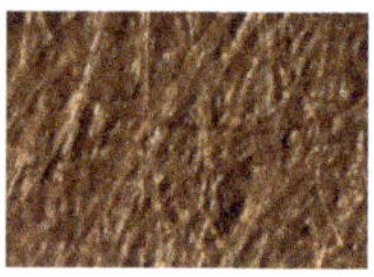	– Brown Kraft paper used for wrapping which has a fairly hard but smooth finish. – Used as a filter paper the brown Kraft paper would trap particles of about 5 µm or greater. – Compared to Whatman paper, Kraft paper had relatively large fibres and was much flatter with a somewhat harder surface.		~5	0.140	0.51
	– Reemay Grade 2250. – Non-woven mat made with spun polyester. – Weight of 19 g/m^2. – It can be used as an air filter.		–	0.127	–
	– Tara linen. – Medium weight fabric (180 g/m^2). – Fine weave of 2.48 yarns per mm in both the warp and fill (weft) directions. – Used in the fill direction as it is stiffer.		–	–	–

pouring it onto a flat and level sheet of polyester film. The concentration used was governed largely by the drying thickness of the film on the polyester. If too thin, the film was too fragile to handle or to be accurately measured for thickness. Increasing the adhesive concentration allowed an increase in the drying thickness of the film. Once initially dried, usually after 30 days or more, the adhesive films were removed from the polyester and cut into sample strips approximately 6–7 mm in width, 150 mm in length and typically 0.10–0.15 mm in thickness.

Peel tests

The basic adhesion test was a simple 'peel' test where the force to separate two adhered sample substrates is measured as illustrated in Figure 3a. The testing equipment was that used for the tensile tests. Since the speed of the peel test is one of the factors influencing adhesion, the tests were conducted at a load rate of 30 second increments between loadings. The displacement rate was 0.0043 cm/sec. All samples were air dried for a minimum of one week prior to testing and tested after equilibrating in the test environments for a minimum of 48

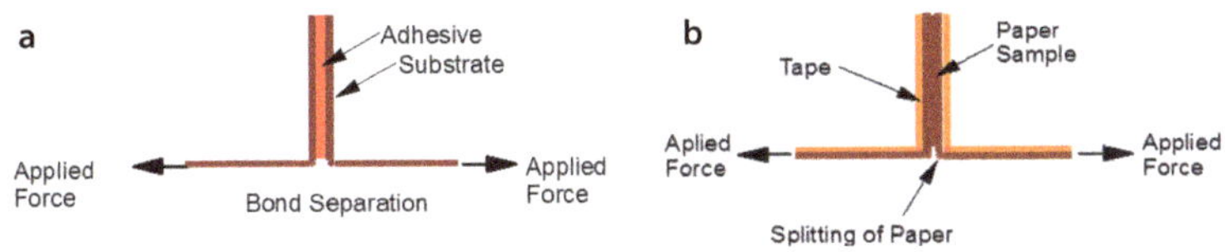

Figure 3 (a) The basic peel test and (b) the use of tape to assist in the splitting of the paper samples.

hours. The RH was controlled using trays of buffered silica gel and monitored using two-stage chilled mirror dew-point meters. Unless otherwise indicated, the test environments were 40–45% RH and 22–23°C.

The substrate materials were typically about 12 mm wide and 125 mm long, and bonded together by submerging them in the adhesive for half of their length and then joining them. Light pressure was used to join them and they were hung and allowed to air dry. For any given sample the length of the bonded substrates was about half of their total length or about 60–70 mm. To start the tests, the samples were peeled apart as far as the bonded section to ensure that only the bonded section was actually measured. For each test a total of about 6.5 mm was peeled apart. To conduct a second test, another section of the sample was used or an entirely new sample was tested. Considering that the substrate surfaces were rather uneven the test results are fairly consistent.

Penetration tests

It was possible to determine the maximum pore dimension that each adhesive was able to penetrate by splitting two papers internally. Figure 3b illustrates the use of tape in assisting in the splitting of the paper samples. Once started, the samples were installed into the micro-tensile testing equipment and the force applied at the same rate as described for the test protocols in the tensile testing section. The force required to separate the papers was measured and recorded in the same manner as that of the adhesive bond test.

Paper splitting controls were determined prior to the application of the adhesives. The range of pore dimensions of this study was 0.0 mm to about 0.025 mm (0–25 μm). For this purpose, different Whatman filter papers (almost 100% cotton and designed to trap particles of certain dimension or larger) and brown Kraft paper (made from pulp fibres) were used. The paper samples were cut in the same direction with dimensions of 6–7 mm in width and 150 mm long.

For testing adhesives, appropriate solutions were prepared and the papers were dipped into the solution for about half their length. Excess adhesive was wiped away and they were hung to dry for a minimum of seven days. Once dry, reinforced packing tape was applied to each side of the sample and the excess tape was cut away from the edges. It was necessary to completely remove the excess tape as any residual tape that stuck to itself at the edges of the sample would have adversely affected the test.

Results and discussion

For this research, natural and synthetic adhesives were tested (Table 1). It should be noted however, that whereas most of them could be tested for adhesion, it was not possible to run cohesive tests for some since no workable free films could be cast. Also, given the length constraints on this paper, results will focus only on some selected natural and synthetic adhesives.

Cohesive testing

This section illustrates the cohesive properties of adhesives, how some change as a function of RH, how solvents affect cohesive properties and how much time they need to evaporate. Figure 4a shows the tensile tests for a 10-year-old hide glue at different levels of RH. In

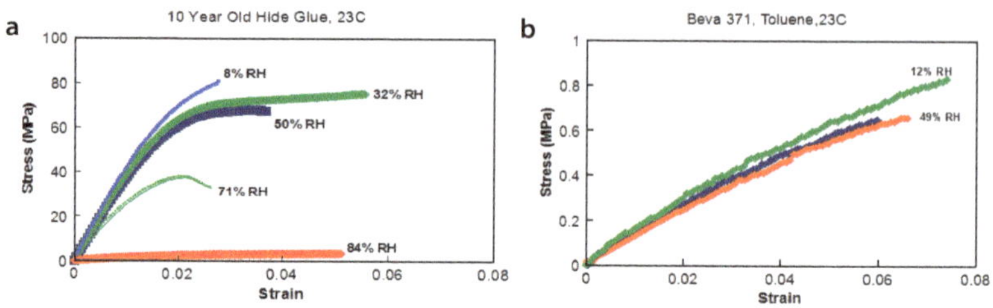

Figure 4 (a) The stress–strain test results of 10-year-old hide glue (10% solution with water and tested at five different RH levels) and (b) the stress–strain test results of Beva 371 (as supplied) and tested at two RH levels.

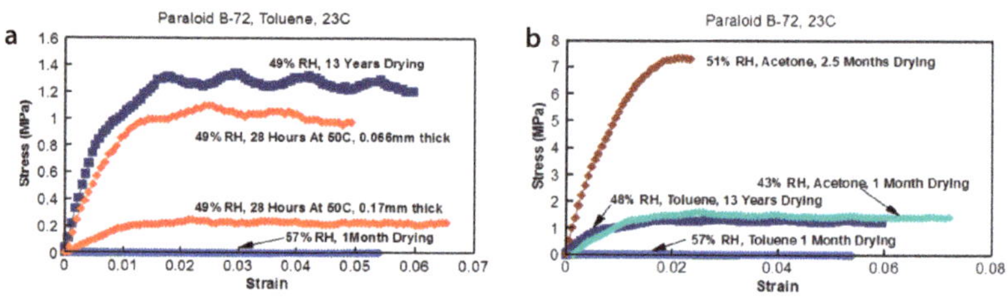

Figure 5 (a) The stress–strain test results of Paraloid B-72 dissolved in toluene after one month, 13 years of drying and being heated to increase the evaporation speed. Different films corresponding to different thickness were also evaluated. (b) The stress–strain test results of Paraloid B-72 dissolved in toluene and acetone after different drying times. In one month the B-72 becomes extremely stiff and brittle.

normal environmental conditions (± 24°C, ± 50% RH), animal glues develop the highest cohesive strengths and stiffness of any of the adhesives tested, far beyond all the synthetic polymers. However, they experience significant changes in their properties as a function of RH. In general, low RH levels make the adhesives become brittle, whereas higher RH (above 80%) cause them to lose nearly all their cohesive properties.

Regarding synthetic adhesives, some are extremely moisture sensitive, such as polyvinyl alcohols, whereas others are not affected by RH, for example, Beva 371 (Figure 4b). The strengths measured vary over two orders of magnitude: from 0.6 MPa in the case of Beva to 68 MPa for animal glues in the same environmental conditions (~50% RH). There are also some adhesives that developed no measurable cohesive strength in normal environmental conditions such as Akeogard T35 and Aquazol 500.

During this research it was also observed that the thickness of the sample had a dramatic effect on its drying time and that in most cases samples needed long drying times before solvents had completely evaporated (thus not influencing testing results). This can be observed in similar stress–strain curves for Paraloid B-72 (Figure 5a). Even after 20 months of drying, Akeogard T35 evidences no cohesive strength at 48% RH. However, it was observed that Akeogard T40 can develop some strength in a relatively short time but still needed around 29 months to allow for full strength development. Similarly, it took 20 months for the water to evaporate from the sample of Klucel G before a reliable stress–strain test could be completed. Independently of the solvent, Plexisol P550 did not develop much strength and still needed considerable time to dry. Interestingly, Aquazol 500 did not present cohesive strength at 52% RH but became brittle at 11% RH. This indicates that lowering the humidity induces a phase change in this material.

From a methodological point of view, it is quite usual to employ volatile solvents in order

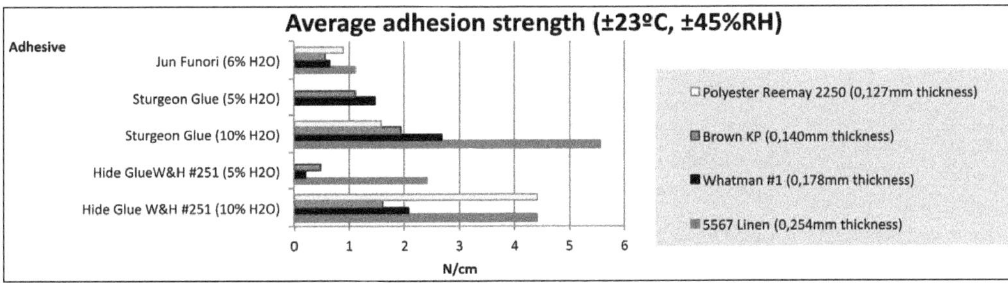

Table 3 The average adhesion strength of some natural adhesives diluted in water in different concentrations as a function of the different substrates bonded.

Adhesive	Solvent and solution	5567 Linen	Brown KP	Whatman #1	Polyester Reemay 2250
Substrate		N/cm	N/cm	N/cm	N/cm
Thickness		0.254 mm	0.140 mm	0.178 mm	0.127 mm
Hide glue W&H #251	10% , H_2O	4.39	1.60	2.06)	4.39
Hide glue W&H #251	5% , H_2O	2.40	0.49	0.19	-
Hide glue W&H #251	10% , H_2O 85% RH	1.60	-	-	-
Sturgeon glue	10% , H_2O	5.57	1.94	2.68	1.58
Sturgeon glue	5% , H_2O	-	1.12	1.47	--
JunFunori	6% , H_2O	1.10	0.56	0.65	0.88

to speed up drying times. The experiments also showed that different solvents can have dramatic effects on the mechanical properties of a single adhesive. For example, Figure 5b shows the tensile test results of Paraloid B-72 dissolved in either toluene or acetone and allowed to dry for differing lengths of time. While a 13-year-old sample of Paraloid B-72 originally dissolved in toluene has a modest strength of around 1.5 MPa, a sample of Paraloid B-72 dissolved in acetone will reach a strength of about 7.5 MPa after drying for only 2½ months. Furthermore, this material when dissolved in acetone becomes extremely brittle.

Adhesion tests

The preceding section was intended to define some properties of the adhesives tested in terms of their stiffness and strength and how these properties change as a result of the solvents used or changes in ambient temperature and RH. The questions posed at the beginning of this paper also dealt with the fact that some of those adhesives that exhibited a lack of strength in the experimental cohesive testing seem to work well in practice. As has been seen, the levels of strength reached for natural adhesives such as animal glue are over a hundred times the strength development of Beva 371 in the same environmental conditions. Why then are both adhesives used to adhere flaking paint with similar apparent success? What needs to be considered to determine whether in both cases the cohesive and/or adhesive requirements are satisfactory?

The literature rarely refers to the strength required to bond fragments of paint to the underlying support and conservators rarely select adhesives based on their strength. Given that cohesive strength is not always a measure of the adhesive properties, adhesive testing is also required. The bar charts in Tables 3 and 4 clearly show the average adhesion strength of many of the adhesives tested (with a variety of concentrations and in different solvents) as a function of the substrates bonded.

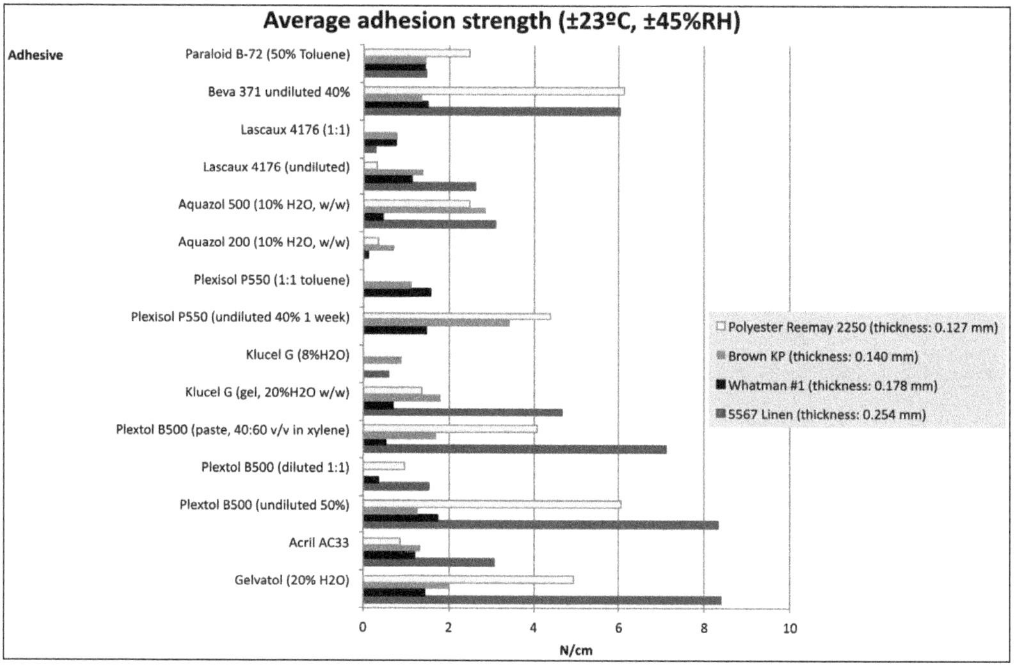

Table 4 The average adhesion strength of some synthetic adhesives diluted in different solvents according to different concentrations as a function of the different substrates bonded.

	5567 Linen (thickness: 0.254 mm)	Brown KP (thickness: 0.140 mm)	Whatman #1 (thickness: 0.178 mm)	Polyester Reemay 2250 (thickness: 0.127 mm)
Gelvatol (20% H_2O)	8.4	2.02	1.43	4.94
Acril AC33	3.06	1.32	1.19	0.86
Plextol B500 (undiluted 50%)	8.32	1.27	1.73	6.06
Plextol B500 (diluted 1:1)	1.55	0.73	0.35	0.96
Plextol B500 (paste, 40:60 v/v in xylene)	7.11	1.7	0.53	4.08
Klucel G (gel, 20%H_2O w/w)	4.67	1.8	0.7	1.36
Klucel G (8%H_2O)	0.58	0.88	–	–
Plexisol P550 (undiluted 40% 1 week)	3.09	3.42	1.45	4.39
Plexisol P550 (1:1 toluene)	0.70	1.13	1.58	–
Aquazol 200 (10% H_2O, w/w)	2.18	0.73	0.11	0.36
Aquazol 500 (10% H_2O, w/w)	3.08	2.85	0.45	2.49
Lascaux 4176 (undiluted)	2.63	1.4	1.14	0.32
Lascaux 4176 (1:1)	0.28	0.8	0.74	–
Beva 371 undiluted 40%	6.01	1.38	1.51	6.13
Micro. wax	0.052	0.17	0.29	0.43
Paraloid B-72 (50% Toluene)	1.45	1.45	1.42	2.5

Hide glue

- *Linen to linen*: the average peel force per width required to separate linen when bonded with a 10% solution of hide glue is 4.39 N/cm. Considering that 1 newton (N) corresponds to a weight of about 102 g, it takes a weight of about 448 g/cm to peel linen bonded with hide glue apart. The peel strength developed for that same sample at 85% RH is about 1.60 N/cm when it should have been expected to be a much lower bond strength given the loss of all cohesive properties at levels of RH above 80% for hide glues. Similarly, the peel force per width for the same linen bonded with the same hide glue but in a 5% concentration is about 2.49 N/cm (slightly more than half of the peel strength of the 10% solution). This reinforces the idea that cohesive tests of an adhesive are not indicative of its bond strength.

- *Linen to polyester*: the average peel strength when using a 10% solution of hide glue to bond linen and polyester mat is identical to the linen even if the tests results show that there is a little more spread in testing. Again, this is surprising since the polyester is non-absorbent and therefore it must be concluded that the bonding mechanism for the hide glue is largely mechanical in that it is surrounding and entangling the textile fibres.
- *Linen to brown Kraft paper*: the force per width needed to split the brown Kraft paper with no adhesive application averages 0.45 N/cm. The 10% hide glue solution and the Kraft paper peel test show average strengths of 1.60 N/cm. This is considerably less than the bond strength of the linen of 4.39 N/cm. A higher paper bond strength might be expected since the paper has a very smooth and porous surface, however the bond strength of the paper is in all probability as high as the linen. In the case of the Kraft paper test with this adhesive (and many others) there were unforeseen results. Considering that the Kraft paper is only 0.140 mm (140 μm) thick, it is significant that the 10% solution is not fully penetrating it. If the adhesive were fully penetrating the internal fibres of the paper, it would require considerably more force to split it apart. On the one hand, the largest pore size of the Kraft paper is estimated to be about 5 μm, which suggests that pore size might be a limiting factor for the penetration of the 10% solution of hide glue. On the other hand, the solvent (water) is rapidly drawn into the paper and away from the adhesive, increasing the adhesive concentration and causing 'blocking'. This has interesting implications since it allows an assessment of the adhesive concentration, depth of penetration of adhesives, and the effect of different substrate pore sizes. It is also noteworthy that the 5% hide glue solution and Kraft paper bond strength is shown to be extremely low. In this situation the original bond line between the papers failed, which was probably the result of 'dry down' where the contraction of the dilute adhesive fails to bridge the gaps between the two paper strips.
- *Linen to Whatman paper*: the Whatman #1 paper is 0.178 mm thick and has a filtering capacity of 11 μm or greater, so this paper is slightly thicker and has greater pore size than the brown Kraft paper. Furthermore, the splitting tests of the Whatman #1 paper with no adhesive gives an average peel strength of 0.43 N/cm, slightly less than the Kraft paper. The peel tests with the 10% solution result in an average splitting strength of 2.06 N/cm (higher than the same glue with Kraft paper) even though this paper is thicker. As with the Kraft paper, the Whatman paper split internally along one of the two paper strips. It is therefore safe to assume that the 10% solution of hide glue is penetrating a little further into the Whatman #1 paper than the Kraft paper. At the 5% glue solution the Whatman #1 bond line fails at a very low level of 0.19.N/cm. Again the glue is drying down during evaporation and not filling the gaps between the two paper strips.

Sturgeon glue

Bonding strips of linen together with a 10% solution of sturgeon glue results in higher (5.57 N/cm) bond strength than the hide glue (4.39 N/cm). Bonding papers together with sturgeon glue has a very similar result when compared to the hide glue and papers. As with the hide glue the papers split internally and

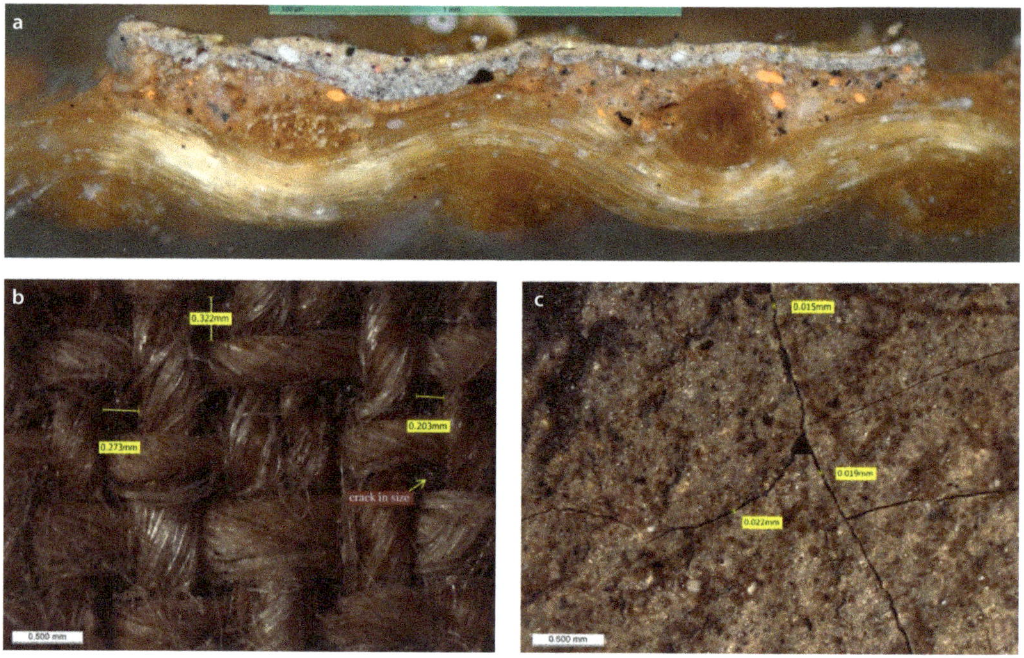

Figure 6 (a) Cross-section of a typical canvas painting; (b) the dimensions of some of the gaps between yarns of a canvas supporting an 18th-century painting; (c) the dimensions of some of the cracks in the paint layer of an 18th-century painting. (Photos courtesy of Melvin J Wachowiak.)

not along the adhesive bond line. Whatman #1 paper showed a higher splitting strength (2.68 N/cm) than the Kraft paper (1.94 N/cm). Again, this is perhaps due to the different pore sizes of the papers and the fact that the bonding strength of the 10% sturgeon glue is higher than the 10% hide glue.

JunFunori

JunFunori is a much weaker adhesive than the hide or sturgeon glue and all tests reflect this. The JunFunori was used in these tests as a 6% solution. The strengths of bonding linen and polyester together with JunFunori are considerably lower than those when using hide or sturgeon glue. Nevertheless the difference between the linen and polyester is small. The average peel strength is very low for both substrates and the linen strength is slightly higher than the polyester. On the other hand, the strength of bonding Whatman #1 and brown Kraft paper together with the 6% solution was very low and the papers split internally, the splitting strength barely higher than the same paper when no adhesive was applied.

Beva 371

In the tests conducted here, the Beva 371 was applied undiluted and no heat was used. The sample substrates were simply pressed together and allowed to dry. The bond strength using the Beva 371 with linen and polyester substrates is very high, even higher than when using hide or sturgeon glue. As with other adhesives, Beva 371 did not fully penetrate the papers and resulted in internal splitting from the tests. However even though the Beva 371 was used undiluted and as a paste it did penetrate the papers sufficiently to develop moderately high splitting strengths.

Aquazol 200/Aquazol 500

Aquazol 200 and 500 together with some of other adhesives tested (Acril AC33, Plextol

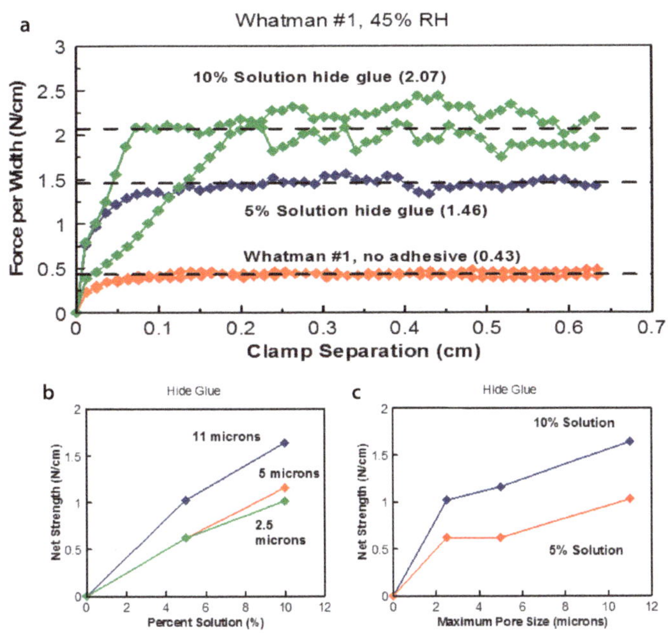

Figure 7 (a) Examples of the test results for splitting Whatman #1 paper with no adhesive and 5% and 10% solutions of hide glue; (b) and (c) summaries of the test results using 5% and 10% solutions of hide glue and the papers with different pore sizes.

B500, Plexisol P550, etc.) showed very little cohesive strength. The peel strength of linen bonded with a 10% solution of Aquazol 500 is slightly more than half that of hide glue and the peel strength of Aquazol 200 is 50% less than the Aquazol 500. Also, and with the exception of the Whatman #1 paper, the bond strength in all cases is fairly good and no internal splitting of either paper with the 10% solution was observed. Even if it is unclear why the Whatman #1 paper did so poorly in the tests, it is worth commenting here that modest adhesive properties can be developed even though there are no cohesive properties for that material. This again suggests that adhesion is largely a mechanical process with most of these adhesives.

Penetration test

Figure 6a shows the cross-section of a typical canvas painting. A scale showing 1 mm can be seen in the micro-photograph indicating that the thickness of a section of the ground and paint layers is only 0.125 mm (125 μm). Accordingly, any cracks, gaps and pores are going to be somewhere in this range. The gaps between the linen yarns of Tara linen 5576 can range between 0.028 mm and 0.212 mm. Older linens tend to have more open weaves with larger gap sizes (Figure 6b). The crack opening dimensions on paint films can range from the very small (0.015 mm) to larger sizes of 0.2 mm and upwards (Figure 6c). The significance is that the gap and pore sizes can vary from 0.001 mm to about 0.5 mm. Actual losses in paint films can become very large.

There are two concerns regarding adhesive penetration:

1. One major concern is whether adhesives will penetrate small gaps and pores with small dimensions in the range from 0.001 mm (1 μm) to about 0.025 mm (25 μm).

2. As the dimensions of pores and gaps get larger there is an additional concern as to the gap-filling capacity of the adhesive.

This section examines different adhesives, their solution concentrations and their ability to penetrate small pore dimensions. At the same time it will also be possible to estimate which adhesive concentrations are blocking, that is, not penetrating the pores at all or failing to fill the gaps in the pores. While Whatman and Kraft papers are not exactly the same thickness they are quite thin and, incidentally, the differences do not become a factor in this experiment. What will be assumed is that the maximum pore size of each paper is approximately that of the largest particle size the papers will retain when used as a filter.

Hide glue

The application of the different solutions of hide glue changes the splitting resistance of the Whatman #1 paper: the stronger the solution, the higher the splitting forces. The test results shown in Figure 7a also exhibit some interesting points. Clearly both solutions of the hide glue are penetrating the 11 μm pores in the paper and the force required to split the paper is actually fairly high. Remembering that the weight of 102 g is about equal to 1 N, the forces shown are the actual results but for the final analyses the paper control forces were subtracted from the measured adhesive–paper forces to give the net influence of applying adhesive to the papers. For example, the force needed to split the paper with the 10% solution of the hide glue is 2.07 N cm whereas the splitting force for the paper alone is 0.43 N/cm. The net increase in splitting force due to the adhesive is therefore 2.07 N/cm minus 0.43 N/cm or 1.64 N/cm. From this point on all results will be discussed in terms of the net force increase due to the adhesive. When using different papers and adhesives different bonding patterns emerge.

For example, Figures 7b and 7c show the results of testing 5% and 10% solutions of hide glue with papers of different pore sizes. In Figure 7b, the net splitting strength of the papers is completely dependent on the solution concentration of the adhesive. For papers of the same pore size the splitting strength is seen to increase with the increased solution concentration. Also shown in this figure is that the increased pore size permits a larger amount of adhesive to penetrate. Consequently the splitting strength is greater for the 11 μm paper than for the 2.5 or 5 μm papers.

Figure 7c shows the splitting strength of the paper versus the pore sizes in greater detail. As with Figure 7b, the lower solution concentration yields lower splitting strengths for each pore size. But of interest here is that as the pore size increases, the splitting does not increase proportionally. At these low pore sizes the adhesive is not fully penetrating but as the pore size increases the bond strength increases somewhat. Keeping in mind that with a 10% solution of hide glue it took about 4.4 N cm to separate two pieces of linen and with a 5% solution it took 2.4 N/cm, the splitting strengths shown for the papers are still well below those values. If papers with greater pore size are used, the strength becomes so high that it cannot be determined, since it defeats the adhesive strength of the tape causing the tape to separate from the paper.

Sturgeon glue

Sturgeon glue has very similar penetrating properties as the hide glue but it shows much stronger splitting strengths. This can be a result of either greater penetration or a simple increase in the strength of the sturgeon glue when compared to the hide glue. Results show that the bond strength is almost proportional to both the pore size and the solution concentration with the highest splitting strength of 2.25 N/cm using the 10% solution and the 11 μm paper. This is an extremely high value and it was not possible to test this concentration of sturgeon glue with papers having greater pore sizes, since the strength of the bond was too high and there was tape failure. Results reveal another aspect of the sturgeon glue relating to the pore sizes discussed with respect to actual artworks. In this figure it shows that when using the 5% solution of sturgeon glue the splitting strength levels off rapidly after a pore size of 11 μm. What is occurring is that this concentration is not gap filling due to the increased pore size and the bond strength shows an upper limit of about 1.1 N/cm. Nearly all the gap and pore dimensions found in damaged works of art are greater than 11 μm and gap filling with a 5% solution of animal glues will present a problem. It will take multiple applications of the adhesive to assure full consolidation. It is also quite possible that with the 10% solution of animal glue, gap filling might be a problem when the pore size is greater than 40 or 50 μm. One further comment here is that the thickness of these papers is about the average thickness of a ground and paint layer found in typical canvas painting and the adhesives under discussion are not fully penetrating the papers.

JunFunori

JunFunori when mixed at a 6% concentration is still somewhat viscous. As a consequence it presents problems in both penetrating and gap filling as it dries down. Results show that initially a small amount of adhesive seems to penetrate using a 3% solution but the splitting force is low. As the concentration is increased there is very little penetration and the adhesive is blocking or left behind on the surface of the paper. On the other hand, as the pore size increases the 6% solution begins to both penetrate and fill the pore gaps. The 3% solution initially penetrates but as the pore size increases, it fails to fill the gaps. The splitting strength of this adhesive is fairly low when compared to either the hide glue or the sturgeon glue.

Beva 371

Beva 371 is a mixture of ethylene vinyl acetate, paraffin wax and ketone resin. Concentrated Beva 371 is reported to be around a 40% solution, but with evaporation it is sometimes not easy to be exact, which is the reason why it is very difficult to determine exact solution concentrations, therefore the information presented below is in dilutions. The tests were performed using a new can of Beva 371 diluted 1:4 with toluene by volume (1 part concentrated adhesive with 4 parts toluene), 1:2 with toluene by volume, and as supplied in its concentrated form (indicated as 1:0). These are not 10%, 20% and 40% solutions. Results show that when using small pore-size papers, it requires high concentrations of Beva 371 to achieve sufficient penetration to develop significant splitting strength. As the pore size of the paper increases, the splitting strength levels out and then decreases. This means that the adhesive is unable to fill gaps at larger pore sizes when using 1:1 and 1:2 dilutions. It is therefore best to use concentrated Beva 371 most of the time when trying to consolidate a painting in a fragile condition. In a way this is somewhat contrary to expectations and it suggests that the Beva 371 mixture is separating and that the lower molecular weight material, possibly the ketone resin, is the only component of the adhesive that is actually penetrating the papers and providing adhesive strength.

Aquazol

At a concentration of about 5%, Aquazol 200 has the ability to penetrate 5 μm pores and can be effective, but it cannot gap fill well at 11 μm pore sizes as shown in Figures 6a and 6b. At a concentration of 2.5% it does poorly in all pore sizes since it does not fill the gaps. For pore sizes greater than 5 μm it will require higher solution concentrations. Aquazol 500 can be a very effective consolidant at 5% solution concentrations in the pore size range from 2.5 μm to 15 μm. For pore sizes greater than 15 μm either higher solution concentrations are required or more than one adhesive application is needed. In general the strength of Aquazol 500 is greater than Aquazol 200.

Paraloid B-72

Paraloid B-72 has been used fairly extensively as a consolidant. For small pore sizes it is reasonably effective at concentrations greater than 10%. It does not appear to penetrate very small pore sizes well. Results show that Paraloid B-72 does not fill voids at a 10% solution and pore sizes greater than 11 μm.

Plextol B500

Plextol B500 penetrates well at the lower concentrations (below 20%) and smaller pore sizes (5–11 μm). At these small pore sizes, higher concentrations tend to block significantly. As the pore size increases the lower concentrations have difficulty filling the voids and gaps. Above pore sizes of 5 μm it will require greater solution concentrations or multiple applications to develop strength. In general this adhesive develops good strength.

Conclusions

The aim of this study was to gain insight into the structural requirements of consolidation adhesives for easel paintings. Some issues regarding the adhesive and cohesive properties of adhesives have been discussed and questions such as penetration, influence of solvents and solvent retention have also been addressed.

As this research has shown, the strengths of the adhesives vary widely in normal environmental conditions (± 50% RH, ± 23°C). Natural adhesives have the highest cohesive strengths and stiffness (modulus) when compared with the other adhesives tested. Synthetic adhesives such as Acril AC33, Plextol B500, Akeogard T35, Aquazol 500 and Aquazol 200 develop no strength or stiffness regardless of how long they are allowed to dry. Many of the adhesives tested develop very little strength or stiffness including Paraloid B-67 in toluene and Plexisol P550 in any solvent tested. With the exceptions of Beva 371 and possibly Paraloid B-72 and Paraloid B-67, the adhesives are dramatically influenced by RH.

For some of the adhesives tested, modest adhesive properties can be developed even though they have scarce or no cohesive strength. Even modest cohesive strength in a material is sufficient to enhance the adhesive properties. This suggests that adhesion is largely a mechanical process with most of these adhesives.

The concentration of the adhesives can also have a significant effect on adhesion: lower concentrations yield lower bond strengths. Critical solution concentrations might exist below which adhesives lose performance and this might also be influenced by porosity. The various types of polymer and their molecular weights are significant factors that require further systematic study. Some of the adhesives tested however appear not to penetrate most of the substrates studied, remaining on the surface and failing to act effectively. This could explain why most of the adhesive and paper tests showed splitting of the paper.

Finally, tests also showed that solvents can take a long time to evaporate from most of the adhesives tested and that the choice of solvent can dramatically affect the mechanical properties of some materials.

Bibliography

Bria C.F. (1986) 'The history of the use of synthetic consolidants and lining adhesives', *WAAC Newsletter* 8(1): 7–11.

Colombo, A., Minotti, D., Mecklenburg, M., Cremonesi, P. and Rossi-Doria, M. (2009) 'Studio delle proprietà meccaniche di consolidanti utilizzati per il restauro di beni policromi mobili', in *The Care of Painted Surfaces: Materials and Methods for Consolidation, and Scientific Methods to Evaluate their Effectiveness*, vol. 2. Padua: Il Prato, 39–48.

Horie, C.V. (2010) *Materials for Conservation: Organic Consolidants, Adhesives and Coatings*. Oxford: Butterworth-Heinemann.

Mecklenburg, M. (2007) *Failure Mechanisms in Canvas Supported Paintings: Approaches for Developing Consolidation Protocols*. Padua: Il Prato.

Mecklenburg, M., Rossi-Doria, M. and Fuster, L. (2008) 'Failure mechanisms in canvas supported paintings: approaches for developing consolidation protocols', in *The Care of Painted Surfaces: Materials and Methods for Consolidation, and Scientific Methods to Evaluate their Effectiveness*, vol. 1. Padua: Il Prato, 49–58.

Paint and Coating Testing Manual (2008) 14th edn, J.V. Koleske (ed), ASTM International.

Authors' addresses

- Marion F. Mecklenburg, Museum Conservation Institute, Smithsonian Institution, Museum Support Center, Room F 2013, 4210 Silver Hill Road, Suitland, MD 20746-2863, USA (mecklenburgm@si.edu)
- Laura Fuster-López, Universidad Politécnica de Valencia, Dept. Conservación y Restauración de Bienes Culturales, Camino de Vera s/n, 46022 Valencia, Spain (laufuslo@crbc.upv.es)
- Silvia Ottolini, Freelance Conservator, via Cima Palone, 16, 43123 Parma, Italy (ottolini.silvia@fastwebnet.it)

The glass transition temperature of adhesives: preliminary guidelines for canvas painting treatments

Christina Young

Introduction

This paper aims to provide preliminary guidelines for the interpretation of glass transition data when assessing the suitability of adhesives for use in a canvas painting treatment. It will attempt to give some guidelines on how to predict physical properties such as stiffness, flexibility and type of fracture that might be expected under different environmental conditions and external forces, such as re-stretching (tensioning). The experimental research that underpins the proposed guidelines was initiated as a result of the frustration at frequently finding insufficient and conflicting data about adhesives' properties when making treatment decisions at the Courtauld Institute of Art's paintings conservation studios. This ongoing research is attempting to establish a methodology for the interpretation of measured thermal transition properties in relation to bulk physical properties. The paper uses a case study to demonstrate how to understand these data and how they may be used.

The majority of adhesives used in the treatment of canvas paintings are viscoelastic. This means that the response of the adhesive to an applied force depends on the temperature, moisture content and the speed of the force; the response can also change over time. The glass transition is the main transition that accompanies a change in modulus (stiffness) from glassy to rubbery properties as the temperature changes. There are four viscoelastic regions – glassy, transition, rubbery and viscous – and all polymers exhibit one or more of these regions. For instance, an adhesive in its glassy region will be stiffer than an adhesive in its rubbery region. This difference can be two orders of magnitude. Some adhesives (especially acrylic and PVAs) have their glass transition region near ambient temperature. This may become more pertinent when considering treatment for paintings entering uncontrolled environments with large temperature ranges or for institutions re-evaluating environmental control strategies. Contemporary or ethnographic paintings in mixed media and/or with additional materials will have viscoelastic constituents and thus have a wide range of mechanical properties that may need to be taken into account, for example in consolidation treatments. Lining of paintings using adhesive mixtures requires knowledge of the transition regions of the combination because more than more one glass transition will occur. By knowing the glass transition temperature (T_g) it is possible to predict properties that affect the structural integrity of the painting such as creep. T_g values can also give a good indication of surface-related properties, such as dirt pick-up, surface texture and degree of gloss, which are relevant in the case of subsequent surface fills and retouching.

Table 1 Adhesives tested.

Adhesive	Type
Resin W interior wood glue (3 months)	Ethylene vinyl acetate
Jade 403N	Vinyl acetate
Jade R	Vinyl acetate
Mowilith DMC427 (3 months)	Vinyl acetate copolymer
BEVA 371 gel (undiluted)	Ethylene vinyl acetate copolymer
BEVA D8	Ethylene vinyl acetate
Aquazol 500 20% wt/v deionised water	Poly(2-ethyl-2-oxazoline)
Plextol B500	Acrylic
Paraloid B-72 25% in Shellsol A	Acrylic

In this paper, glass transition data have been compared with uniaxial tensile data that give the bulk properties of the adhesive, for example stiffness, strength and mode of fracture. The relationship between the T_g and the mode of failure (ductile or brittle fracture) is important, because if a brittle fracture occurs, it is more likely to lead to damage of the surrounding canvas or paint. However, bulk testing of adhesives over transition regions is very time consuming and, in some cases, hard to achieve if the adhesive has a very low transition temperature.

Transition temperatures are measured by thermo analysis techniques, primarily differential scanning calorimetry (DSC) and dynamic mechanical analysis (DMA). Both techniques are explained in the appendix. Data available to the conservator are often confusing because no single value uniquely describes T_g; this is exacerbated as the method of measurement is often not given. However, it is well established within polymer mechanics that the T_g is dependent on the thermo analysis technique used and interpretation of the data (Chartoff 1997). Different definitions are used depending on the method of measurement and in what context the data will be used. This is because, as stated earlier, the way the force is applied will affect the way the material responds, rather than an error in methodology. This can lead to very different predictions of the physical properties of the materials and hence the correct choice of adhesive for a conservation treatment. The numerical differences that can occur have also been demonstrated for a subset of resins used in conservation (Schilling 1989).

Samples

The adhesives chosen for testing are used in canvas painting conservation and represent generic types used within the profession (Young 2011). These were polyvinyl acetates (PVACs), Bostik Findley Interior Wood Glue Resin W, Jade 403N and Jade R, ethylene vinyl acetates Beva 371 and Beva D8, poly (2-ethyl-2-oxazoline) Aquazol 500, acrylic-based adhesives Plextol B500 and Paraloid B-72 (Table 1).

Free films of the adhesives were drawn down onto flexible polytetrafluoroethene (PTFE) sheets taped to rigid boards. The wet film thickness was controlled by two parallel lines of PVC tape, four layers thick (each layer 0.12 mm), along which a knife edge was drawn. They were made in a controlled environment (relative humidity (RH) 50% ± 2% RH and a temperature of 20°C ± 2°C) then left uncovered in ultraviolet (UV) filtered light for one week to allow any volatile components to evaporate. Subsequently, the samples were placed in airtight opaque boxes with silica gel matting (Artsorb) conditioned to 55% RH and stored in ambient conditions (typically 55% ± 10% RH, 20°C ± 4°C). Once a week the boxes were opened to allow any further volatile components to escape. The films were left for a minimum of six weeks to ensure that they had reached their dry film thickness. To minimise any pre-strain, the flexible PTFE was peeled away from the sample, rather than peeling the sample film. All test samples were cut from the same films and were nominally 6 mm wide, 0.4 mm thick, with a gauge length of 5 mm for both the DMA and static uniaxial tensile testing. Prior to testing, the width was measured with Vernier calipers with an experimental error of ± 0.01 mm, and the thickness was measured with a micrometer with an experimental error of ± 0.001 mm.

Static uniaxial tensile testing

Static uniaxial tensile testing was carried out on an Instron 4301 with an integrated environmental chamber. The load, extension, RH and temperature were logged throughout the testing. Samples were clamped in the upper grip and aligned, then allowed to come into equilibrium at the desired conditions (typically 20 minutes) before tightening the bottom grip. A crosshead speed of 5 mm/min was used because this lies between the average speeds of the DMA test runs 1 Hz (1.2 mm/min) and 10 Hz (12 mm/min). It also represents a speed at which induced strains might occur on works of art from processes involving tensioning (not including shock) or environmentally induced deformations (Young and Ackroyd 2001). A minimum of three samples was tested at each test temperature and RH.

Differential scanning calorimetry (DSC) and dynamic mechanical analysis (DMA)

Thermo analysis was carried out at the Getty Conservation Centre, Los Angeles. Three of each sample type were tested for both DSC and DMA measurements. Measurements with the Mettler Toledo DSC 822 followed a method, developed by Schilling (1989), for the measurement of polymers such as Paraloid B-72 (see the appendix). After baking out the DSC, an In/Zn blank pan was run to provide a baseline measurement from which the thermal characteristics of the test materials could be extracted. Nitrogen was used as the purge gas, discs were cut from the free films and placed in the sample pan ensuring good thermal contact. The typical weight of the sample was 2–4 mg. The sample weight was measured prior to testing using a micro balance to an accuracy of ± 0.1mg.

DMA was carried out on a Triton 2000. In dynamic mode the instrument can operate in a variety of configurations – tension, shear, cantilever bend and three-point bend (Duncan 2008). Most operating parameters are user defined. Previous testing on acrylic and alkyd paint films had allowed pre-selection of the best combination of parameters to obtain meaningful results. It was found that, for the type of adhesive samples tested, single cantilever gave the most repeatable results. Samples were tested at three frequencies: 1 Hz, 3.16 Hz and 10 Hz. All samples were run on the DSC first to obtain a T_g value to ensure that the DMA temperature scan covered this region.

Results

Interpretation of DSC curves

The DSC output data is a measurement of heat flow; the flow of energy into or out of the adhesive as a function of temperature or time. Figure 1a shows the output data for the complete heating DSC cycle as specified in the appendix for Resin W. A distinct transition can be seen between 10°C and 20°C. It is this portion of the data that is used to calculate the T_g. Figure 1b shows this region in detail and illustrates how T_g is determined. First a tangent line is drawn from the near flat horizontal part of the curve; a second tangent line is then drawn from the near vertical part of the curve. The point at which these two lines intersect is taken as the T_g onset. (An alternative method takes the midpoint on the vertical tangent line – this is called the T_g midpoint). For this work the T_g onset has been used because (as will be shown later) this corresponds to the start of the change in mechanical properties with temperature. The two curves shown in Figure 1b are for the first and second heating cycle of the adhesive. This process of heating twice removes any volatile components, which is thought to better represent the condition of the cured adhesive and produce a more reproducible result. Although beyond the scope of this paper to discuss fully, it is worth mentioning that for unstable adhesives or slow curing adhesives, the first cycle T_g onset value may be

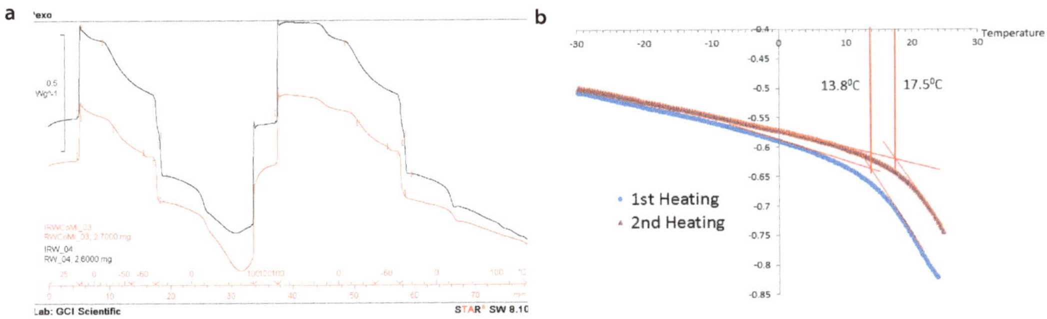

Figure 1 (a) DSC output for Resin W showing the glass transition region (first and second heating cycle) corresponding to the method in the appendix; (b) DSC output for Resin W showing a detail of the glass transition region (first and second heating cycle) and the derivation of the T_g onset.

relevant. This factor needs further investigation. Taking the convention of the second cycle for T_g onset for Resin W, the value is found to be 17.5 ± 0.6°C.

Interpretation of DMA curves

Figure 2 shows the DMA storage modulus and tan δ curves for Resin W at three different frequencies (1 Hz, 3.16 Hz and 10 Hz). The shift towards higher temperatures with increasing test frequency for the storage modulus and tan δ curves has been discussed in other publications and will not be repeated here (Chartoff *et al.* 2009; Young 2011). It is a result of the strain rate dependency of viscoelastic

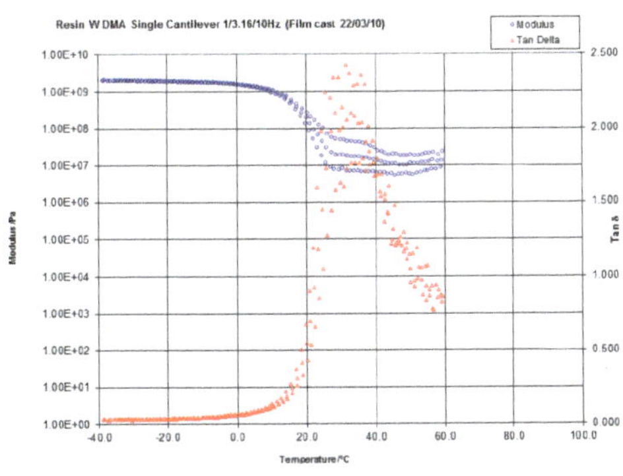

Figure 2 DMA storage modulus and tan δ curves for Resin W.

polymers, and thus it is important that comparisons in T_g between adhesives are made at the same frequency or at least the shift should be taken into account. The curve on the left corresponds to the change in stiffness (storage modulus) as a function of temperature. It can be seen that between approximately 10°C and 30°C this modulus changes by two orders of magnitude. To ascertain the onset of this change the intersection of the horizontal tangent and vertical tangent should be taken, which for this sample is 14.2°C. Also shown in Figure 2 is the tan δ curve (see the appendix). The temperature at which the tan δ peak occurs is often used as the point to obtain the T_g, in this case 31.2°C. Therefore it can be seen that there is a wide range in what can be defined as the T_g for this adhesive. However, because of the way each definition of the glass transition is derived, usually the tan δ value is the highest and the DSC onset the lowest. Interestingly, the DSC onset from first heating gives the lowest value equal to 13.8°C. This is closest to the DMA storage modulus onset value. In both cases, the samples had not been previously heated, were tested within hours of each other and were from the same sample batch. Thus the assumption can be made that any volatile components remaining in the sample will be the same and that these two values represent the same state of the sample. If a direct comparison with the second heating DSC value is required then the

Table 2 T_g values measured from the DSC second heating cycle.

Adhesive	T_{1g} (midpoint)[1]	T_g (onset)
Resin W interior wood glue (3 months)	22	17.5
Jade 403N	8.2	1.6
Jade R	-7.2	-12
Mowilith DMC427 (3 months)	20	7
Beva 371 gel (undiluted)	-25	-30
Beva D8	11	7
Aquazol 500 20% wt/v deionised water	40	32
Plextol B500 (no thickener added)	3.4	-12.6
Paraloid B-72 25% in Shellsol A	5	-3
Araldite 1253 (Young 2011)	–	43.8

DMA testing first includes a pre-heating cycle. Some published DSC data have included this in the test methodology (Gabbott 2008). The relative merit of each method is an area for further research, especially for non-stable adhesives such Resin W. However, what is of most interest here is how thermomechanical values allow the bulk tensile properties to be predicted, properties that relate directly to the physical state in which the adhesive will be on the painting.

Interpretation of uniaxial tensile curves

Figure 3 shows a graph of the bulk tensile stiffness versus %strain at three different temperatures for DMC427. At +15°C the adhesive is in its rubbery state, at -5°C it is in its glassy state with little plastic deformation before fracture (brittle), and at +5°C the adhesive is still in its glassy state but with some plastic deformation (line starting to curve) before fracture (ductile). The corresponding DSC onset value (on second heating) for the adhesive was measured at +7°C (Table 2).

The value of the brittle–ductile transition therefore occurs close to the DSC onset value. For adhesives that are in or close to the rubbery region at room temperature, obtaining accurate and repeatable data using DMA or bulk tensile testing is very difficult experimentally. So far it has not been possible to confirm this relationship for all the adhesives tested here. This correlation between the brittle–ductile transition and the T_g, as measured by the DSC onset, has been shown to hold for other adhesives including epoxy Araldite SV1253 and filled Resin W (both used in the UK for panel painting conservation) and acrylic and PVA primers. Full details of these results are published elsewhere (Young and Hagan 2008; Young 2011). Thus, at this stage based on all the data available it is reasonable to suggest that the relationship is valid.

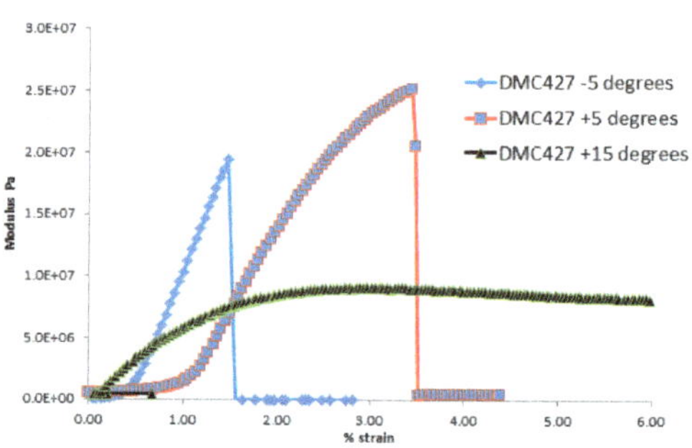

Figure 3 Uniaxial tensile stiffness versus %strain at three different temperatures for DMC427.

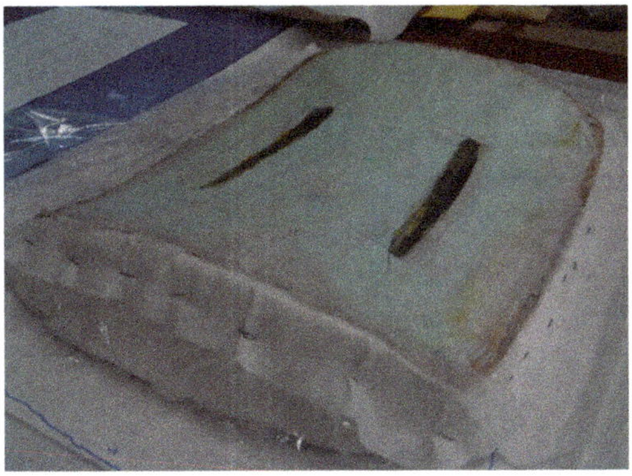

Figure 4 Ron Gorchov, untitled painting.

Case study using available data

An untitled painting by Ron Gorchov from 1947 treated in the Conservation and Technology Department at the Courtauld Institute of Art in London demonstrates the use of T_g data. It also shows how to use T_g data when, as is often the case, only partial information is available. The painting support (25 × 40 cm) was made from a cotton/muslin canvas stretched over a saddle-shaped stretcher with compound angles (Figure 4). The painting had suffered an impact or had been dropped resulting in a broken stretcher member, severe distortion of the canvas and cracking of the paint layers; the canvas had become detached from two edges. Consolidation of the paint layers was required as these layers were going to be put under high strain as the canvas reacquired its original saddle shape when tensioned. Retaining flexibility within the layer structure was paramount while the painting was returned to its original shape using localised application of moisture and heat, and subsequently on re-stretching. Empirical tests on the painting found that the best results for handling, penetration and paint adhesion were achieved most successfully with Aquazol 200, and reasonably well with DMC427. Both adhesives are moisture sensitive. DMC427 has been found to increase in stiffness over time but does not result in brittle failure in ambient conditions (Young 2011). The results from the DSC data give Aquazol 500 T_g onset 32°C and DMC427 T_g onset 7°C. Although Aquazol 200 has a lower molecular weight than Aquazol 500 the manufacturer states that they both have a T_g between 69 and 71°C (it is not clearly stated how this is measured). This high value compared to the DSC onset value suggests that it is derived from DMA tan δ (previous testing has shown the difference is typically 25°C (Young and Hagan 2008)). If the T_g is not significantly varying with molecular weight then the DSC onset value of 32°C can be used for the Aquazol 200. Because this value corresponds to the ductile–brittle transition it predicts that Aquazol 200 would be brittle around ambient temperatures and therefore it was not suitable for this treatment (empirically this adhesive when cast as a film feels brittle at room temperature). Therefore, for the Ron Gorchov painting, DMC427 was chosen as the consolidating adhesive.

Conclusions

This paper has attempted to demonstrate (using Resin W, DM427 and Aquazol 500 as examples) why different glass transition temperatures can be given for the same material.

Table 3 Measurement options for mechanical properties.

Mechanical property	Measurement data
Strength	Bulk tensile data
Stiffness	Tensile data or DMA storage modulus (approx.)
Mode of failure (ductile/brittle)	Bulk tensile data or DSC T_g onset to predict
Will it creep?	DMA curve to find transition region or DSC data to predict*
Will it be tacky?	DMA curve to find rubbery region or DSC data to predict*

*Other mechanical tests can measure these properties directly but the data do not exist for most adhesives used in conservation. However, simple empirical tests can be conducted to gain a feel for these properties at room temperature.

It has also shown how it relates to the ductile–brittle transition and how this may be used to help in the choice of a suitable adhesive for a particular treatment.

Some more general guidelines may also be given based on the data presented in this paper and in previous publications:

- DSC onset data is more reproducible – it usually has values closest to the ductile–brittle transition and lowest value for T_g.
- DMA tan δ gives the highest value for T_g.
- DMA tan δ is at least 15°C higher than the start of transition in stiffness.
- DMA tan δ is at least 25°C higher than the ductile–brittle transition.

Table 3 suggests which data would be most useful when choosing an adhesive, depending on which mechanical property is considered most relevant at a certain temperature.

Appendix: Principles of DSC and DMA

A DMA instrument applies an oscillating force (stress σ) to a sample and measures the resultant dimension change (strain ε). It is measured under controlled temperature (scanning or isothermal). For a perfectly elastic material the stress and strain are perfectly in phase. DMA measures the amplitudes of the stress and strain, as well as the phase angle (δ) between them. This value is used to resolve the modulus into the storage modulus E' and the loss modulus E". These moduli are simply related to Young's modulus E (calculated from the initial linear gradient of the stress/strain curve as measured using static uniaxial tensile testing) by $E^2 = (E')^2 + (E'')^2$. Tan δ = loss modulus/ storage modulus.

A viscous material has a high value of tan δ whereas an elastic material has a low tan δ (Duncan 2008).

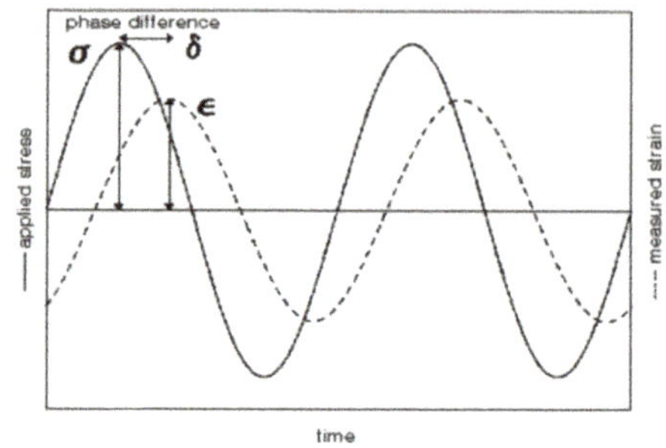

DSC

A DSC instrument measures the difference between the heat flow from the sample and a well-characterised reference material as the temperature is changed. Changes in heat capacity during a glass transition are determined from the difference in heat flow (Gabbott 2008).

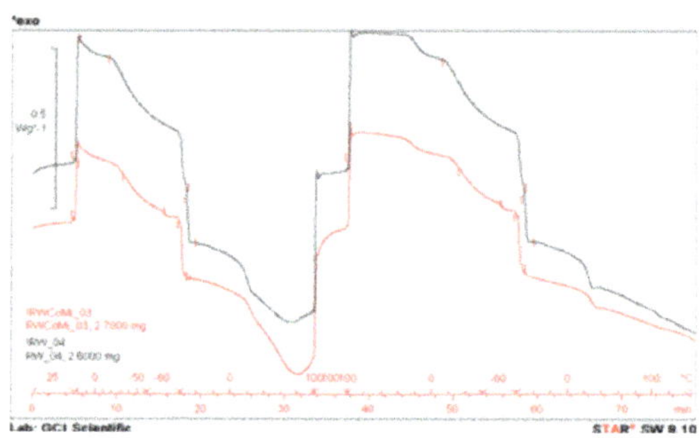

DSC method used for the tests (Schilling 1989)

1. 25°C 5 min
2. 25–60°C at 10/min
3. 60°C 4 min
4. 60–100°C at 10/min
5. 100°C 4 min.
6. 100–60°C at 10/min
7. 60°C 4 min
8. 60–150°C at 10/min

All with N_2 80ml/min

Materials and suppliers

- Jade R/Jade 403N/ Aquazol 500/ Beva D8: Talas, Morgan Ave Brooklyn, NY 11211, USA, www.talasonline.com/.
- Beva 371: AP Fitzpatrick, 142 Cambridge Heath Road, Bethnal Green, London, E1 5QJ, www.apfitzpatrick.co.uk.
- Resin W interior wood adhesive: Axminster Tools, Trafalgar Way, Axminster, Devon EX13 5PB, UK, www.axminster.co.uk.
- Araldite Epoxy 1253: Conservation Resources (UK), Ltd, Unit 2, Ashville Way, Off Watlington Road, Cowley, Oxford OX4 6TU, UK, www.conservation-resources.co.uk/.

Acknowledgements

I would like to thank the following: the Getty Foundation and the Getty Conservation Institute (GCI) for funding and hosting me as a Visiting Scholar under which this work was initiated; Michael Schilling and Alan Phenix (Getty Conservation Centre); John Duncan and Glynn Van-de-Velde (Triton Technology Ltd); Prof. Tony Kinloch (Imperial College of Science, Technology and Medicine); Amy Griffin (Courtauld Institute of Art).

References

Chartoff, R.P. (1997) 'Thermoplastics', in E.A. Turi (ed.), *Thermal Characterization of Polymeric Materials*. San Diego: Academic Press, 483–743.

Chartoff, R.P., Menczel, J.P. and Dillman, S.H. (2009) 'Dynamic mechanical analysis', in J.D. Menczel and R.B. Prime (eds), *Dynamic Mechanical Analysis in Thermal Analysis of Polymers*, 2nd edn. Hoboken, NJ: Wiley.

Duncan, J. (2008) 'Principles and applications of mechanical thermal analysis', in P. Gabbott (ed.), *Principles and Applications of Thermal Analysis*, 2nd edn. Oxford: Blackwell, 119–89.

Gabbott, P. (2008) 'A practical introduction to differential scanning calorimetry', in P. Gabbott (ed.), *Principles and Applications of Thermal Analysis*, 2nd edn. Oxford: Blackwell, 1–50.

Schilling, M.R. (1989) 'The glass transition of materials used in conservation', *Studies in Conservation* 34(3): 110–16.

Young, C.R.T. (2011) 'Predicting practical properties of unfilled and filled adhesives from thermo-mechanical data', in CCI *Symposium – Adhesives and Consolidants for Conservation: Research and Applications Preprints*, Ottawa. Available at: www.cci-icc.gc.ca/symposium/2011/Paper%2012%20-%20Young%20-%20English.pdf (accessed 3 September 2009).

Young, C.R.T. and Ackroyd, P. (2001) 'The change in the mechanical response of easel paintings to three typical lining treatments', *National Gallery Technical Bulletin* 22: 85–104.

Young, C.R.T. and Hagan, E. (2008) 'Cold temperature effects of modern paints used for priming flexible supports', in J.H. Townsend, T. Doherty, G. Heydenreich and J. Ridge (eds), *Preparation for Painting: The Artist's Choice and its Consequences*. London: Archetype Publications, 172–9.

Young, C.R.T., New, B. and Marchant, R. (2011) 'Experimental evaluation of adhesive gap filler combinations for joining panel paintings', *Facing the Challenges of Panel Paintings Conservation: Trends, Treatments, and Training*. Los Angeles: The Getty Conservation Institute,125–39.

Author's address

Christina Young (Young@courtauld.ac.uk)

Lining with a fixed interleaf: a case study in the structural effects of paper interleaf and adhesives

Cecil Krarup Andersen and Ingelise Nielsen

Introduction

The purpose of this research was to assess the degree of structural support offered by aged linings, with and without a fixed interleaf, at a range of different ambient humidity levels. When using a fixed interleaf it is attached to the lining canvas before the lining process, usually with an adhesive that is different from the lining adhesive. Historically interleaves have been used for several purposes and they are still a part of the paintings conservator's toolbox, but the effects that additional layers in a lining have on structural support and on moisture sensitivity of paintings had not been examined.

In the 1960s the use of an interleaf was introduced as part of the lining process in an attempt to reduce the problem of the canvas structure becoming visible in the paint layer during the lining process, a phenomenon commonly known as 'weave interference'. The problem was caused by heat and pressure when lining on the new hot tables. An interleaf of synthetic non-woven fabric, inserted between the lining canvas and the original canvas, was proposed by Linard as a possible solution (Linard 1965).

The use of interleaves became increasingly popular among conservators during the 1960s–70s. A questionnaire from 1974 suggested that interleaves were used by most conservators (Rees Jones *et al.* 1975) and that paper was the most commonly used interleaf material, followed by felted synthetic fibres. In 1984 a successive questionnaire showed that interleaves were then mainly used in the USA (Hedley and Villers 1984) and that a wide variety of materials was in use. When a third survey was made in 2002 (Ackroyd *et al.* 2002) interleaves were found to be used mostly in Australia, Austria, the Netherlands, Germany and Brazil. The most frequently used interleaves were now paper, woven gauze-like fabrics and Mylar (a polyester film).

The reasons for using interleaves were to reduce texture change in the original, to improve reversibility and to increase stiffness (Ackroyd *et al.* 2002) as research had shown that stiff lining materials would reduce the risk of further cracks and cupping in a painting and better support repaired tears in the painting canvas (Mecklenburg 1982; Hedley 1981). However the question remained: how stiff is stiff enough?

To be successful in minimising cracks, cupping and paint loss, the lining structure must be stiffer than the paint (Hedley 1981; Mecklenburg 1982). Elm (1953) showed conservators (Keck 1969) that the strain at failure in an aged paint layer can be as little as 1%. Mecklenburg confirmed this result and showed that some types of white paint crack already at a strain of 0.5% (Mecklenburg 2005).

Therefore it can be concluded that a paint layer should not be subjected to more than 0.5% strain to avoid risk of failure or deformation of the paint.

The next question then is: what are the typical forces when paintings are tensioned? There have been several attempts to define the force with which a canvas painting is usually tensioned. Authors use different definitions of force that are here converted to N/cm. Hedley assessed that the normal load put on a painting was 2–4 kg/inch measured uniaxially (Hedley 1981). This can be converted to approximately 7.7–15.4 N/cm. Berger stated that a level of 200 N/m (2.0 N/cm) was a normal tension for a newly stretched canvas painting whereas 60–130 N/m (0.6–1.3 N/cm) biaxially measured was the maximal level that most canvases can hold once they are tensioned on a fixed stretcher (Berger and Russell 1988). Young and Hibberd (1999) argued that 30–60 N/25 cm (1.2–2.4 N/cm) would be a normal range for a newly stretched or lined paintings measured biaxially.

It can be argued therefore that a material that does not exceed 0.5% strain when stretched 15.4 N/cm uniaxially in both weave directions could be considered stiff enough to serve as a lining canvas (the tests in this study were made uniaxially). High tensile strength and a minimum of creep are also desirable properties for a lining material. In order to improve stiffness, conservators have experimented with different lining materials but due to crimp in the yarns none of the widely used lining fabrics seem to be stiff enough at the abovementioned force levels.

Berger's studies of the use of interleaves found that using Mylar as an interleaf could make the linings stiffer and create a moisture barrier on the back of the paintings (Berger and Russell 2000: 91). He also asserted that adhesion between the layers of a lining was enhanced by the use of an interleaf (Berger 1966).

The method, referred to as fixed interleaf, was examined and refined in Denmark during the 1970s –80s as part of lining treatments using a low-pressure vacuum table (Bondesen and Ketnath 1973; Ketnath and Bøgh 1975). In this lining tradition interleaves were attached to the lining canvas with an adhesive different to the lining adhesive (Scharff 1995). The use of a secondary adhesive secured full control of the lining process and prevented the interleaf from becoming distorted or moving out of plane during the lining. The purpose of the fixed interleaf was to provide strength, stiffness and reversibility to the lining. Reversibility would be achieved because the paper, if thick enough, could be split down the middle with minimal use of force when the lining was peeled off.

Since interleaves have been used frequently in the past it is important for conservators to be able to assess the effects of interleaves and additional layers of adhesive. In this research lined samples were tested to investigate the amount of support achieved at different relative humidity (RH) levels and the reversibility of this lining method.

The tested samples

Four different naturally aged lined samples were tested. Three of the linings had different paper interleaves and the fourth was a control lining with no interleaf. The control sample had three layers: a primed canvas to resemble a painting, Plextol D360 adhesive and Ulster linen as the lining canvas. Plextol D360 has been identified on the aged samples by Fourier transform infrared spectroscopy-attenuated total reflectance (FTIR-ATR). Plextol D360 is an aqueous dispersion of a butyl acrylate and methyl methacrylate-based copolymer. Its minimal film-forming temperature is 0°C, which means that the adhesive will remain tacky at room temperature. After more than 20 years this was still true in the case of the test samples. Acrylic resin dispersions have been used for linings since the 1970s especially in Scandinavia, France (Ackroyd *et al.* 2002) and the Netherlands (Mehra 1972).

The three lined samples with interleaves had two extra layers in the lining: a paper interleaf and a secondary adhesive that was identified by FTIR-ATR as polyvinyl acetate (PVAc) and

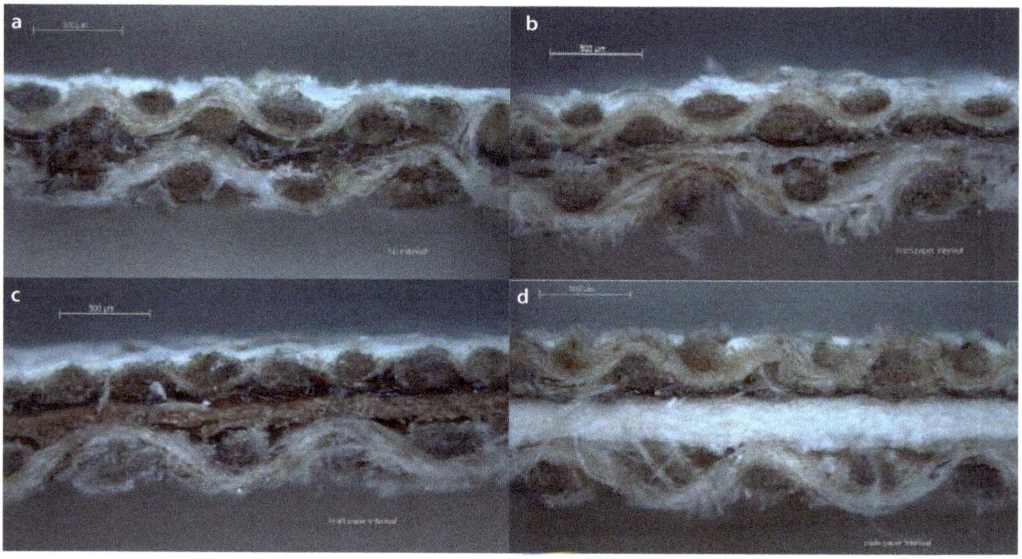

Figure 1 Cross-sections of naturally aged lining samples: (a) no interleaf (lining canvas, lining adhesive and primed canvas); (b) kozo paper interleaf (lining canvas, lining adhesive, paper, secondary adhesive and primed canvas); (c) Kraft paper interleaf (lining canvas, lining adhesive, paper, secondary adhesive and primed canvas); (d) plate paper interleaf (lining canvas, lining adhesive, paper, secondary adhesive and primed canvas).

methyl cellulose. The mixture of PVAc and methyl cellulose has been used in conservation because it combines the strength of the PVAc with the water solubility and high viscosity of the methyl cellulose. The methyl cellulose facilitates the application of PVAc with a brush as it prolongs the drying time and makes the adhesive smoother. PVAc became widely used in conservation in the 1950s, but a few decades later concerns were raised about its use because of poor ageing qualities and the fact that it can be hard to remove once it is dry (Horie 1987: 94–7). As an emulsion the adhesive is more unstable and additives such as plasticisers are often added to avoid brittleness. As it ages PVAc becomes stronger and has also been said to become more acidic and release acidic vapour (Down *et al.* 1994). However, different kinds of PVAc have different properties at different times during ageing, depending on their specific method of manufacture (Horie 1987: 10). Thus recent research detected no acidity in 30-year-old PVAc linings of posters; in fact the linings had prevented acidity in the posters (Recasens 2010). Some kinds of PVAc are still used occasionally for linings of paintings (Ackroyd *et al.* 2002) but in Denmark this use is rare now.

The types of paper interleaf used in the aged samples were identified as kozo, Kraft paper and plate paper by FTIR-ATR and microscopic fibre analysis. The four test samples are shown as cross-sections in Figure 1.

Since the aged samples could not be taken apart to test the materials individually, reference materials were used to investigate the behaviour of the individual layers. Rolls of the same canvas and papers used in the aged samples have been kept at the School of Conservation at the Royal Danish Academy of Fine Arts, Copenhagen, for decades. This made it possible to determine which materials were used to construct the lined samples and the identification was confirmed by FTIR-ATR and microscopic analysis. The reference materials are the three types of paper, the two types of adhesive and the lining canvas.

In the aged samples the weave directions of the lining canvas and the primed canvas, used to represent a painting, were parallel. They

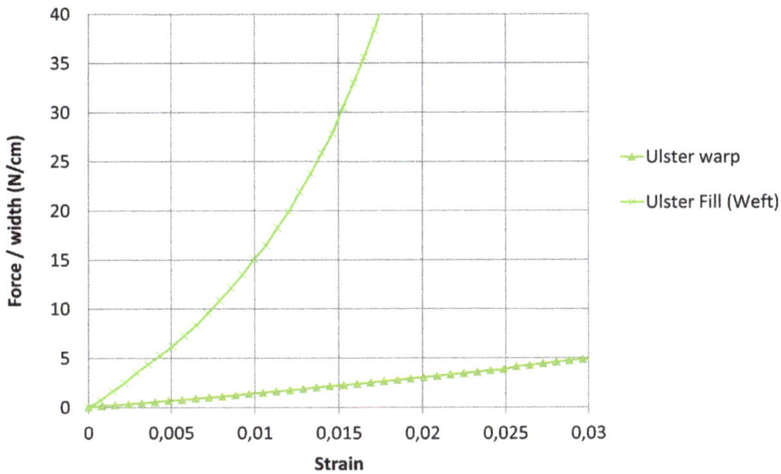

Figure 2 Force strain curves for the lining canvas reference material in both weave directions.

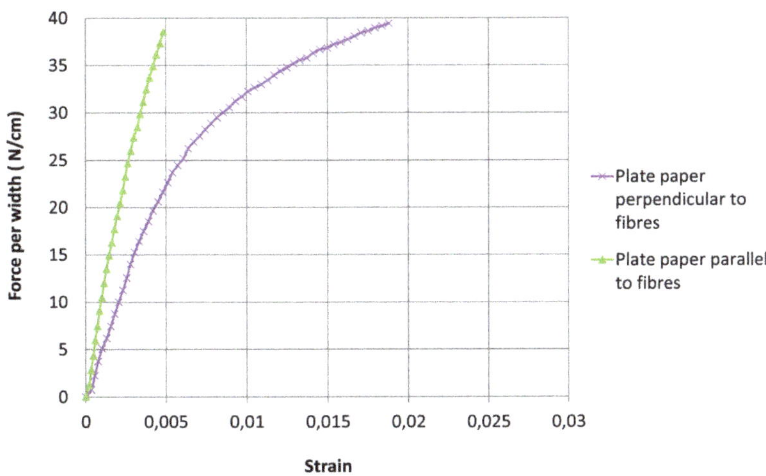

Figure 3 Force strain curves for plate paper reference material in both fibre directions.

could, however, only be tested in the fill (weft) direction because of the way they had been constructed. The reference samples could be tested in two directions: the canvases in warp and fill directions and the papers in fibre direction and cross-fibre direction.

Methods

Force governs failure in a material, and cracks and cuppings are therefore less likely to occur if the forces acting on the paint film are reduced. In order to determine if a lining is actually being effective in reducing stress in a paint layer, a stress–strain test can be carried out that will show which layers are stiffer and more supportive in a lined painting.

Stress is defined as force per cross-sectional area, but since it is inconvenient to measure the cross-sectional area of a canvas, force per width is mostly used for research in paintings conservation. The force per width/ strain curve shows what happens in a painting when it is stretched on a stretcher or keyed out. Both

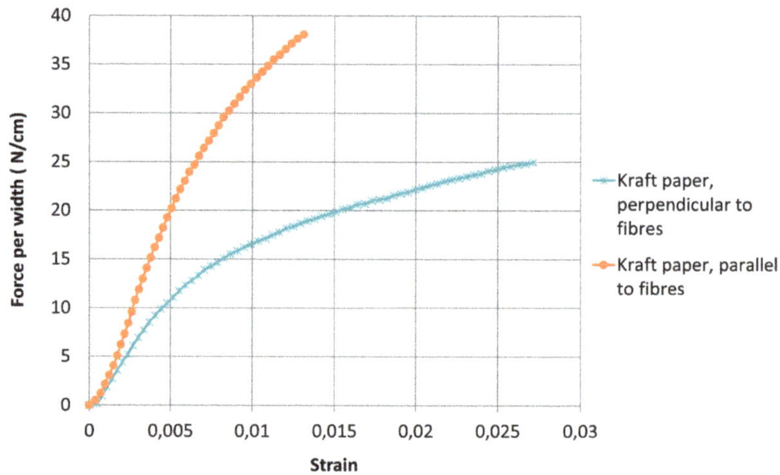

Figure 4 Force strain curves for Kraft paper reference material in both fibre directions.

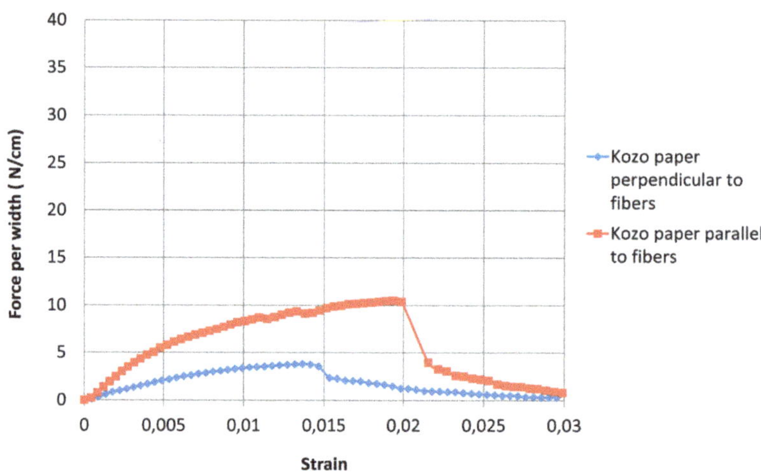

Figure 5 Force strain curves for kozo paper reference material in both fibre directions.

these scenarios can induce enough force in paint layers to make them to fail.

Force per width/ strain tests were done to determine the stiffness of the samples. The samples were 10 × 140 mm (gauge length 100 mm) and the displacement speed was 0.137 mm per 30 seconds. Before testing the samples were conditioned to approximately 50% RH and 20°C for a minimum of 18 hours to help ensure equilibrium with the environment in which they were tested.

A restrained specimen test was used to investigate the reaction of a tensioned painting to fluctuating RH. The test was performed in a humidity chamber where the sample was tensioned to 2.5 N/cm. After the first stress relaxation the sample started to react in a repeatable pattern to the cycled RH and the changes in force per width were measured as the RH changed. The RH in the chamber was changed stepwise using silica gel or saturated salts. When the sample reached equilibrium with the air in the chamber after a minimum of 18 hours, the change in force per width was recorded and the moisture level was changed again.

A T-peel test (American Society for Testing

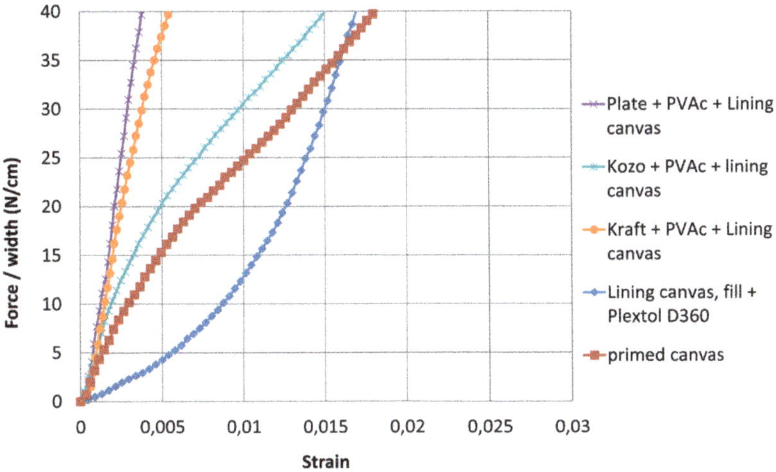

Figure 6 Force strain curves for lining materials and primed canvas.

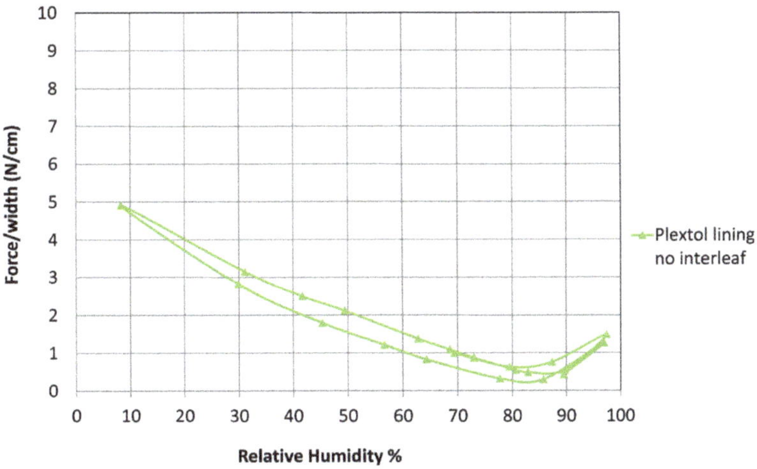

Figure 7 Force / relative humidity curve for lining with no interleaf.

and Materials (ASTM) D 1876-00) for reversibility was used to determine how much force was needed to peel the lining canvas off the original canvas. The tensile tester used was an Instron 5566. The sample size was 25 × 190 mm, the initial gauge length was 50 mm and the speed was 254 mm/minute.

Results

Force per width/ strain tests of the separate layers in the lining provided an understanding of the role played by each layer: painting, lining adhesives, interleaves and canvas. The reference materials were tested to understand how each material in a lining adds to the structural behaviour of the lining at 50% RH.

The lining fabric reference, Ulster linen, showed significant differences in stiffness between fill and warp (Figure 2). The fill direction was stiffer and therefore provided more support to the painting than the warp direction. When the three interleaf papers were tested they showed different stiffness levels (Figures 3–5). The plate paper added significantly to the overall stiffness of the lining when used as an interleaf whereas the kozo

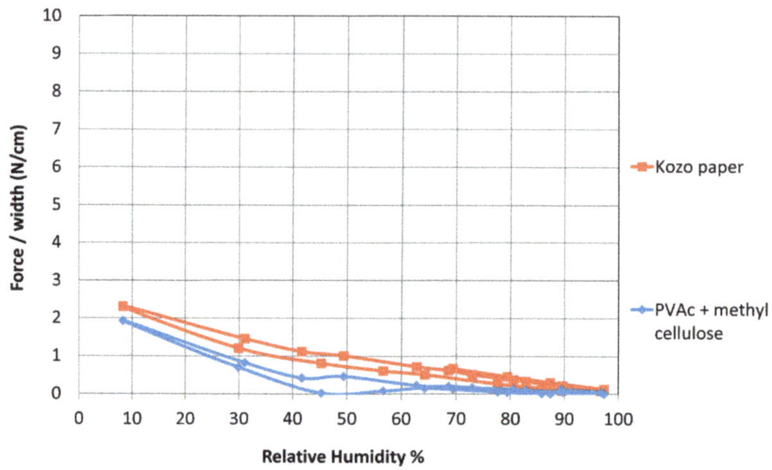

Figure 8 Force / relative humidity test of reference samples, kozo paper and PVAc + methyl cellulose.

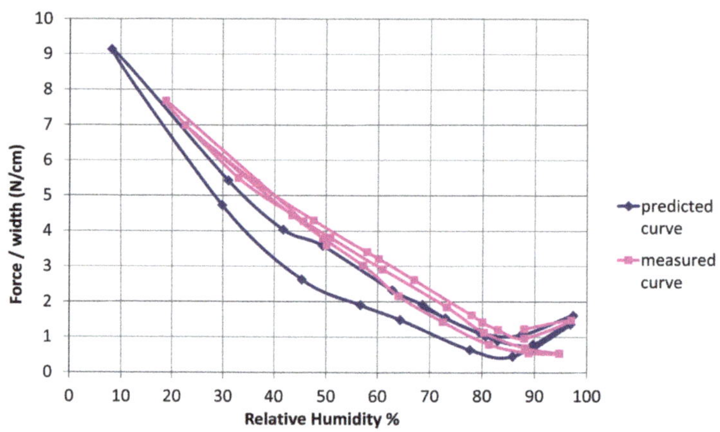

Figure 9 Predicted curve for the lined sample with kozo paper interleaf and the measured curve.

paper was much more flexible. Orientation of the fibres in the interleaf papers resulted in different stiffnesses depending on the direction tested.

The two adhesives in the lining samples had very different properties. By looking at the combined stiffness of the linings with different types of fixed interleaf it became evident that a significant amount of stiffness originated from the aged PVAc and methyl cellulose adhesive due to cross-linking in the aged adhesive (Figure 6).

The Plextol D360 remained tacky after more than 20 years due to the low glass transition temperature. It therefore had a very low cohesive strength and hardly provided any stiffness to the structure as evidenced by the fact that the lining canvas with Plextol D360 and no interleaf had more or less the same force strain curve as the raw canvas (Figures 2 and 6). The stiffness of the combined lining materials is compared to the stiffness of the primed canvas in Figure 6. Kozo paper did not add as much to the stiffness of the lining as the two thicker and stronger types of paper. The lining with the kozo paper interleaf was still stiffer than the primed canvas in the measured fill direction

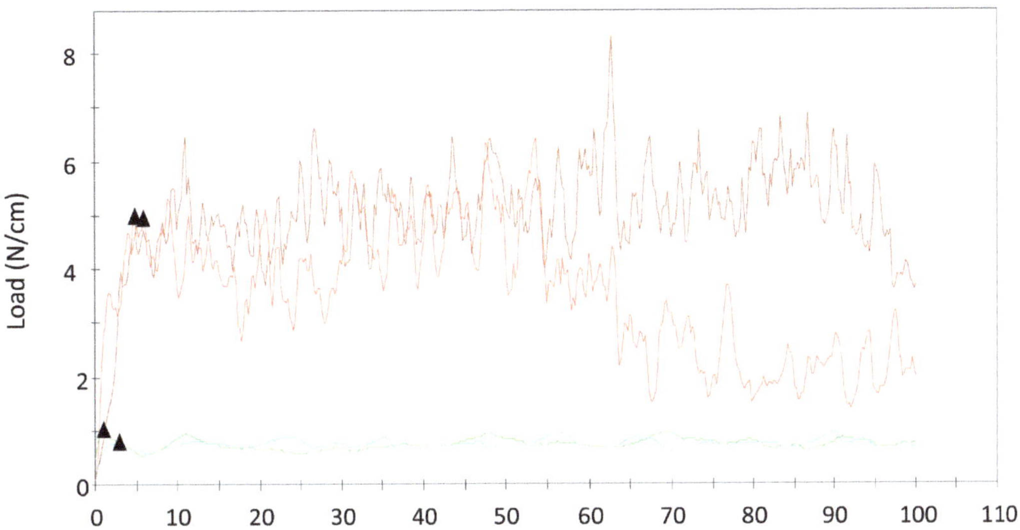

	Average Load 20-100 mm (N/cm)
1: PlextolD360 lining -No interleaf	3,42
2: PlextolD360 lining with Kozo paper interleaf	5,23
3: PlextolD360 lining with Kraft paper interleaf	0,73
4: PlextolD360 lining with Plate paper interleaf	0,74

Figure 10 T-peel test of naturally aged lining samples.

and would therefore be able to support the primed canvas and avoid cracks, cupping and tears at around 50% RH. However, much of this support seemed to come from the aged PVAc + methyl cellulose as the kozo paper alone was not very stiff in any direction. The canvas with Plextol D360 alone was not nearly as stiff as the primed canvas and would not be able to support the primed canvas and avoid damage originating from force applied at 50% RH. The next section investigates how this picture changes with variations in the ambient moisture levels.

Response to moisture

Hygroscopic materials change tensile properties with fluctuating RH. Mecklenburg (1982) and Hedley (1988) demonstrated how each layer of a tensioned painting influenced its behaviour in changing RH. The curves they obtained for canvas paintings illustrated how paintings contract in both very high and low humidity. Figure 7 shows the primed canvas lined with Plextol D360 with no interleaves. This curve is very similar to the response to RH that Mecklenburg and Hedley found for untreated paintings.

The increasing force that can be seen in the layers of the lined, primed canvas during desiccation at low RH is caused by the loss of moisture in restrained materials, in particular the glue size in the primed canvas. At 85% RH the forces in the lined painting approach zero and in very humid conditions the two canvases try to contract due to swelling/ shrinkage in the two canvases and possible casting strains from the size layer. From this curve it can be deduced that the stiffness of the lined painting is much higher at low RH than at 85% RH.

So, what happens to the curve when a fixed interleaf is added? Each layer in the lining has a different moisture coefficient of expansion and as such responds differently to changes in RH. The resulting force in the restrained and lined primed canvas is governed by the principle of superposition. This principle states that when a composite material is subjected to an external load, for example caused by a change in RH or tensioning, the resultant forces are equal to the sum of the forces in the separate members provided that all of the materials are maintained at the same degree of restraint (Hearn 1997: 27). When measured forces in the individual layers of a lined painting are added it is therefore possible to predict the behaviour of the lined painting. The total force in the lined painting will be equal to the sum of the forces in each material. For any given moisture and strain level the following therefore applies:

$$F_{(\text{lined painting})} = F_{(\text{painting})} + F_{(\text{lining adhesive})} + F_{(\text{interleaf})} + F_{(\text{secondary adhesive})} + F_{(\text{lining fabric})}$$

The assumption is that the force that develops in each restrained material does not affect the results of the other materials provided that all the materials are maintained at the same degree of restraint. This means that each material can be tested separately and the combined effect on the painting calculated.

Figure 8 shows the 'force versus RH' plot of two interleaf materials: one is an adhesive mixture of PVAc and methyl cellulose, 1:2 (w/w), and the other is kozo paper. This mixture of PVAc and methyl cellulose is an approximation as the exact recipe is not available. Age can also be an important factor in the reaction to RH as cross-linking is likely to make the adhesive less hygroscopic in time. Consequently, this curve for the new adhesive mixture will probably represent the maximum effect that can be expected from this mixture of adhesives.

During desiccation both materials attempt to contract but as they are restrained and cannot move forces develop. At high RH levels the materials expand and thereby lose their restraint. As the adhesive mixture takes up moisture it loses strength and bonding ability at levels above 60% RH, and the other materials in the lining become free to move both in and out of plane.

When extra layers of interleaf materials are included in the lining the forces that develop at all RH levels can be totalled using superposition as shown earlier. The forces developed in the plot shown in Figure 9 represent the influence of the combined forces in the primed canvas with Plextol D360 lining (Figure 7) and the kozo paper interleaf and secondary adhesive (Figure 8). At low humidity, the combination of the restrained materials results in high forces in the lined painting. Figure 9 shows both the measured curve and the predicted curve, where the principle of superposition has been used to predict the effect that the two extra layers in the lining have on the total force development.

The curve showing the force reaction to fluctuating RH is considerably steeper than before the extra layers were applied. It can therefore be concluded that the paper interleaf and the extra hygroscopic adhesive add to the build-up of force and thereby to the potential risk of failure and deformation of the paint layers. The forces in the lining structure have almost doubled.

Reversibility

Another factor that may cause damage in lined paintings is the adhesion between the painting and the lining materials. According to Phenix and Hedley (1984), poor adhesion between layers may cause delamination and out-of-plane deformation and they therefore set a requirement for minimum lining peel strength of 300 g/2.5 cm, which translates into 1.18 N/cm. If this requirement is not met the authors claim there is a risk of delamination and out-of-plane distortions.

One of the arguments for using the thicker papers as interleaves is that reversibility is increased because the paper can split down the middle when the lining canvas is removed from the original canvas. This argument was tested in a peel test (Figure 10) and the result confirmed that in the two samples with thicker paper interleaves, less force was required to separate the lining canvas from the original canvas. The kozo paper is so thin that the glue penetrated it and reinforced the binding, making it difficult to split this paper.

However, a drawback of the splitting of the paper was that after the peel test of the interleaf samples, paper fibres and Plextol D360 were left on the original canvas. An attempt was made to remove the fibres and Plextol D360 by softening the matrix gently with heat from a hot air blower. While heating, the adhesive and fibres were scraped off with a scalpel. It was possible to remove the fibres and Plextol D360 in all three cases but the process was time consuming and a fragile original canvas could potentially suffer from such a treatment. It is not clear if this would be more damaging than peeling off a Plextol D360 lining without the interleaf.

The sample with no interleaf left no visible adhesive or fibres in the surface of the original canvas after the peel test but removing the lining took more force and thereby added more stress to the original canvas. Therefore, the effort needed to remove paper and adhesive from the original canvas should be taken into account when assessing the reversibility of the lining methods. Many lining adhesives would present the same challenges when removed and, as with so many other conservation treatments, we can hardly claim absolute reversibility.

Conclusions

At 50% RH the samples with an interleaf were very successful in stiffening the structure, due to both the interleaf and the additional adhesive but the interleaf samples became significantly more responsive to fluctuating RH. The lining without an interleaf offered a less effective support but was also less responsive to fluctuating RH.

Since the introduction of nap bond linings, adhesives have rarely been considered when assessing the stiffness of a lining even though they can add significant stiffness depending on their cohesive strength. Adhesives with high cohesive strength and stiff interleaf materials are possible ways of improving lining techniques. In fact the tests suggested that with interleaves it is possible to produce linings with tensile properties that can hinder failure and loss in brittle paint layers. The low cohesive strength of the thicker interleaf papers makes them easy to split and thereby improves reversibility.

The drawbacks are that the low cohesive strength in the paper may also result in unintended delamination of the lining or out-of-plane distortion of the painting depending on the magnitude of shear forces between the layers in the lining.

The two properties – reversibility and stiffness – sometimes seem to work against each other. Stiff linings can be difficult to remove and linings that can be easily removed are often not stiff enough to prevent cracks in the paint layers. Linings with an interleaf and a stiff secondary adhesive seem to point towards a solution for these problems provided that the chosen interleaf materials are not responsive to moisture.

This research has also shown how support and response to humidity in a lining can be predicted by testing the layers individually. The influence that a lining has on a painting's moisture sensibility can therefore be evaluated and become part of the transport, storage and conservation plans in paintings collections. By testing materials and using the principle of superposition it is possible to determine which paintings need additional protection and care because their linings make them more prone to force-induced damage.

If a painting is lined with more layers of hygroscopic material it will result in increased moisture sensitivity and risk of delamination as well as failure and loss in the paint layers.

Paintings with hygroscopic interleaf layers require extra attention when subjected to changing environmental factors. However, non-hygroscopic canvases, interleaves and adhesives present a promising possibility for devising linings that support paintings and at the same time restrain their response to environmental changes.

Material specifications

Primed canvas

- *Bast fibres*: 27 fill threads/cm and 28 warp threads/cm. The samples are cut in the fill direction.

Paper

- *Kozo paper*: kozo paper is a Japanese paper made of fibres from the bark of the paper mulberry tree (*Broussonetia papyrifera*). Kozo fibres are long and strong. The paper is unsized and lignin-free. Approx. thickness: 0.11 mm. Approx. weight: 28 g/m^2.
- *Kraft paper*: Kraft paper is a brown paper made from Kraft (sulphate) pulp. It is a strong and coarse paper used primarily for wrapping or packaging purposes. The paper contains lignin and has a smooth and a rough side. Approx. thickness: 0.11 mm. Approx. weight: 79 g/m^2.
- *Plate paper*: plate paper is a thick printing paper with a smooth surface. The paper is sized and lignin-free. Approx. thickness: 0.22 mm. Approx. weight: 147 g/m^2.

Adhesive

- *PVAc + methyl cellulose*: the recipe for the PVAc-based adhesive in the lined samples is unknown but the reference was made by mixing 2 volume units methyl cellulose 3.5% and 1 volume unit PVAc dispersion (Cascol carpenter glue). This recipe was based on good application properties for lining. This reference had dried for one month when the force per width/ strain tests were done.

Lining canvas

- *Ulster linen canvas made from flax fibres*: 22.5 warp threads/cm and 16.0 fill threads/cm. 223,63 g/m^2. The samples were all cut in the fill direction. The samples were conditioned for a minimum of 12 hours before testing to ensure equilibrium with the environment in which they were tested.

Acknowledgements

The authors wish to thank Marion F. Mecklenburg, Senior Research Scientist at the Smithsonian Institution and Mikkel Scharff, Head of Department at the School of Conservation, The Royal Danish Academy of Fine Arts for help and advice.

References

Ackroyd, P., Phenix, A., Villers, C. and Wade, N. (2002) 'Structural treatment for canvas paintings in 2002: summary of questionnaire replies', in *Preprints of ICOM-CC 13th Triennial Meeting*. London: James & James, 321–7.

Berger, G.A. (1966) 'Weave interference in vacuum lining of pictures', *Studies in Conservation* 11: s.170–80.

Berger, G.A. and Russell, W.H. (1988) 'An evaluation of the preparation of canvas paintings using stress measurements', *Studies in Conservation* 33: 187–204.

Berger, G.A. and Russell, W.H. (2000) *Conservation of Paintings: Research and Innovations*. London: Archetype Publications, 85–108.

Bondesen, P. and Ketnath, A. (1973) 'Gabriel Engels: *David and Bathsheba* (1649): restoration and survival', *Meddelelser om Konservering* 2(2): 33–45.

Down J., MacDonald, M.A., Tétrault, J. And Williams, R. Scott (1994) *Adhesive Testing at the Canadian Conservation Institute: An Evaluation of Selected Poly(vinyl acetate) And Acrylic Adhesives*, Environment and Deterioration Report No. 1603. Ottawa: Canadian Conservation Institute.

Elm, A.C. (1953) 'Some mechanical properties of paint films', *Official Digest* 346: 751–74.

Hearn, E.J. (1997) *Mechanics of Materials 1. An Introduction to the Mechanics of Elastic and Plastic Deformation of Solids and Structural Materials*, 3rd edn. Oxford: Butterworth-Heinemann.

Hedley, G. (1981) 'The stiffness of lining fabrics: theoretical and practical considerations', in *Preprints of ICOM-CC 6th Triennial Meeting*. Paris: ICOM, 81/2/2-1-13.

Hedley, G. (1988) 'Relative humidity and the stress/strain response of canvas paintings: uniaxial measurements of naturally aged samples', *Studies in Conservation* 33: 138–42.

Hedley, G. and Villers, C. (1984) 'Lining in 1984: questionnaire', in D. de Froment (ed.), *Preprints of ICOM-CC 7th Triennial* Meeting. Paris: ICOM in association with the J. Paul Getty Trust, 84.2.22–25.

Horie, C.V. (1987) *Materials for Conservation: Organic Consolidants, Adhesives and Coatings*. Oxford: Butterworth-Heinemann.

Keck, S. (1969) 'Mechanical alteration of the paint film', *Studies in Conservation* 14: 9–30.

Ketnath, A. and Bøgh, L. (1975) 'Om anvendelsen af EP 1 dispersion ved doublering'/ (('About the use of EP 1 dispersion for lining'), *Meddelelser om Konservering* 2(5): 144–51.

Linard, H. (1965) 'Amélioration, dans certains cas, du procédé de rentoilage à la cire-résine par l'emploi de "someline"', *Studies in Conservation* 10(1): 24–6.

Mecklenburg, M.F. (1982) *Some Aspects of the Mechanical Behaviour of Fabric Supported Paintings*, Report to the Smithsonian Institution (unpublished).

Mecklenburg, M.F. (2005) 'The structure of canvas supported paintings', in *Preprints of the Interim Meeting of the International Conference on Painting Conservation. Canvases: Behaviour, Deterioration and Treatment, Paraninfo de la Universidad Politecnica de Valencia, 9–11 March 2005*, 119–55.

Mehra, V.R. (1972) *Comparative Study of Conventional Relining Methods and Materials and Research towards their Improvement*. Brussels: ICOM Committee for the Care of Paintings.

Phenix, A and Hedley, G. (1984) 'Lining without heat or moisture', in *Preprints of ICOM-CC 7th Triennial Meeting*. Paris: ICOM in association with the J. Paul Getty Trust, 84.2.38–39.

Recasens, C.R. (2010) *Evaluación de los efectos causados por el uso de PVAc en los entelados de cartels modernos / Evaluation of the Effects Caused by the Use of PVAc in Lining Modern Posters*, PhD thesis, Universitat de Barcelona, Departemento de Pintura, Facultad de Belles Artes.

Rees-Jones, S., Cummings, A. and Hedley, G. (1975) 'Relining materials and techniques: summary of replies to a questionnaire', *Preprints of ICOM-CC 7th Triennial Meeting*. Paris: ICOM, 751131-4.

Scharff, M. (1995) 'The development and present status of structural conservation treatments of canvas paintings in Denmark', in *Lining and Backing: The Support of Paintings, Paper and Textiles. Papers delivered at the UKIC Conference, 7–8 November 1995*. London: United Kingdom Institute for Conservation of Historic & Artistic Works, 48–53.

Young, C.R.T. and Hibberd, R.D. (1999) 'Biaxial tensile testing of paintings on canvas', *Studies in Conservation* 44(2): 129–41.

Authors' addresses

- Cecil Krarup Andersen,The Royal Danish Academy of Fine Arts, Schools of Architecture, Design and Conservation, Esplanaden 34, 1263 Copenhagen K, Denmark (cka@kadk.dk)
- Ingelise Nielsen, The Royal Danish Academy of Fine Arts, Schools of Architecture, Design and Conservation, Esplanaden 34, 1263 Copenhagen K, Denmark (in@kadk.dk)

Teaching adhesive principles to conservation students: *sine scientia ars nihil est* – without knowledge, skill is nothing[1]

Kate Seymour

Introduction

This paper highlights the methods used to teach adhesive principles to first-year conservation students from diverse backgrounds in a multidisciplinary context. Adhesives are for the most part polymers and thus are key components in the conservation practitioner's toolbox. These are often not only the 'building blocks' of the materials requiring conservation, but also the 'tools' with which a conservator will work to repair, reconstitute or restore damaged or degraded works of art. More importantly, because these materials are frequently introduced into original structures, it is essential that the conservation student should have a thorough and wide knowledge and understanding of the potential adhesives available. The range and properties of polymeric materials used in conservation is introduced and the physical and chemical properties of both fresh and aged material are considered within the content of the module. Principles of reversibility and long-term performance of adhesives used in different applications are also discussed.

Adhesives are required to bridge the gap between two (divergent) substrates, bond to both adherends and remain stable, physically and chemically, on solidification (Allen 1984; Berger and Zeliger 1984). They are able to become a liquid either by dissolving in a solvent, by phase change on heating, or are in liquid form prior to a chemical reaction. The choice of specific adhesives is of central importance in relation to a particular task or project. Numerous adhesive formulations are available commercially and specific preparations can easily be made in the conservation laboratory. Manufactured products contain a number of unknown components that may make their behaviour in the long term unpredictable. This applies particularly in combination with other materials where reversibility may be an issue to consider. Self-made formulae are easy to engineer and by definition contain specifically selected known materials (Wolbers 2008). It remains up to the conservator to decide which adhesive to use and at which concentration for each specific task. Teaching students the skills to make ethical selections must be based on a thorough understanding of the materials available and their ageing characteristics. Students must be able to compare materials by considering properties such as viscosity, wettability, diffusion, flexibility, bond strength, glass transition temperature and degradation (Horie 2010). They can be encouraged to collect the relevant data for each adhesive, however visualising this data without practical experience is moot.

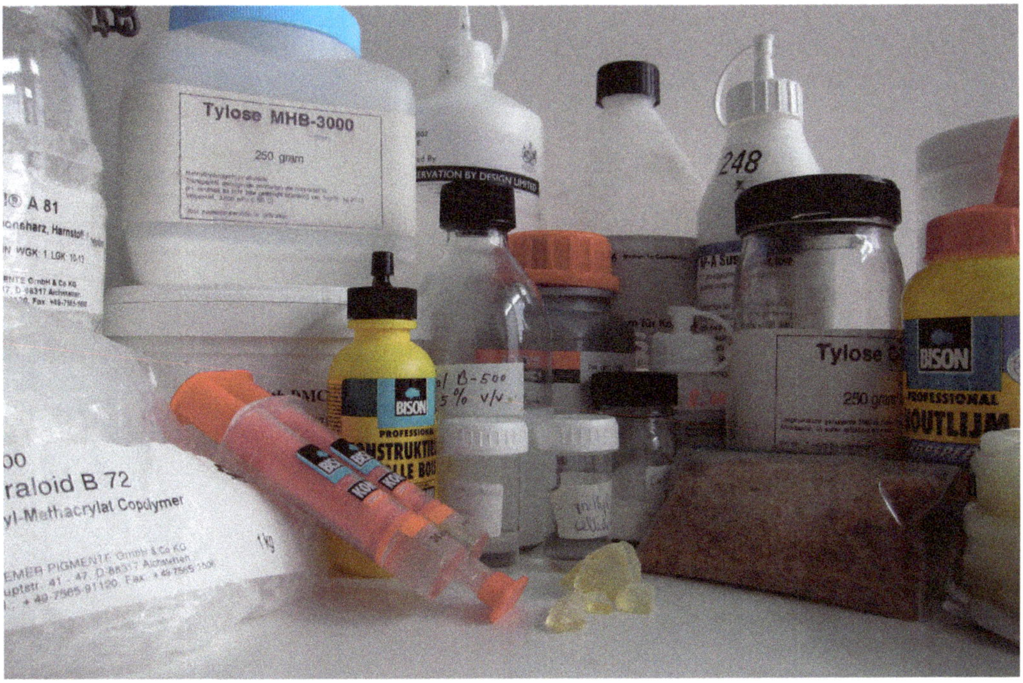

Figure 1 Which adhesive? A challenging question!

Back to basics: domains of competence

Most adhesives used within the conservation field are polymers or form polymers on curing. These can be either natural or synthetic in origin. Adhesives can be categorised according to their chemical components, their curing system (phase change), or their function. Manuals, which outline the theory of adhesion, are produced and updated regularly. Criteria governing adhesive characterisation, technology, function and degradation are widely reported in publications. Evaluation of a range of parameters, including bond strength and joint failure of individual adhesives, has been described both within our field and outside it. Furthermore, a few new and improved polymers have been proposed within the conservation field.[2] The choice of adhesive for a particular task remains a challenging question that may not have a single answer. Thus, what is the essential knowledge that should be taught and how is that information best provided to enable a novice to become a 'competent practitioner' (Keck 1977)?[3]

Three key factors enable the student to make informed decisions, in turn learning which adhesives are appropriate for a particular function or situation. First, a broad chemical knowledge of adhesives commonly used within the conservation community is indispensable. An understanding of the physical properties of these adhesives follows logically and lastly, an understanding of adhesive–surface interactions forms the third leg of this particular three-legged stool.[4] The domains of competence that a student should conquer include material knowledge, communication skills, practical skills and problem solving. Thus, it is essential that theoretical knowledge be underpinned with practical examples and real-life scenarios. Students are stimulated to discuss these experiences and are given challenging situations to solve. It is the pedagogical approach, rather than the principles themselves that will be discussed further in this paper.

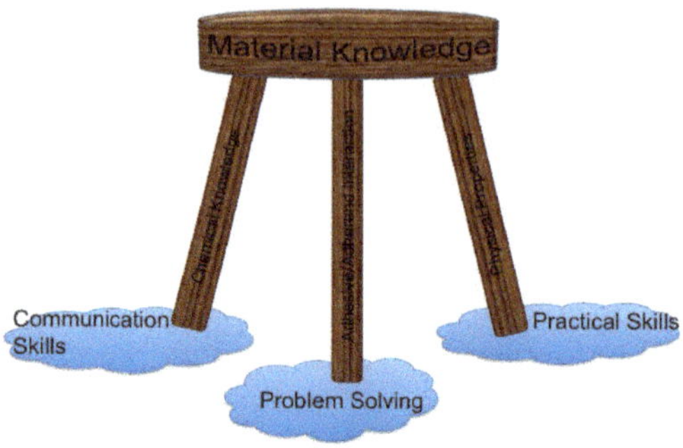

Figure 2 Back to basics: domains of competence.

Student demographics

The Conservation and Restoration of Cultural Heritage programme at the University of Amsterdam began in 2006 and the first graduates received their diplomas in 2010. While this university is a newcomer to the league of educational institutions offering this subject at a tertiary level, the formation of the course has a longer history as the curriculum was derived through the amalgamation of two pre-existing programmes: those offered at the Instituut Centraal Nederland (a four-year higher professional education course) and at the Stichting Restauratie Atelier Limburg (a five-year post-academic course). Disciplines offered by both institutes were distinct and now form the eight specialisations available at the University of Amsterdam: Book and Paper, Modern and Contemporary Art, Glass, Stone and Ceramics, Textiles, Easel Paintings, Wood and Furniture, Metals and Historic Interiors. This new course is designed, within the spirit of the Bologna process, to fulfill requirements outlined by the European Confederation of Conservator-Restorers (ECCO) and was ratified by the European Network for Conservation-Restoration Education (ENCoRE) in 2008. However, the Dutch accredited course is unique as it offers a seven-year programme consisting of a three-year Bachelor, a two-year Masters and a two-year postgraduate phase. Introduction modules to conservation history and ethics, materials, preventive conservation and conservation science are taught within a general Bachelor programme in either science or humanities. Selected students elect one of the eight disciplines offered at the beginning of the MA programme and will only graduate as conservators if all four subsequent years are completed.

The demographics of students entering the MA level conservation programme at the University of Amsterdam show that most have studied arts and languages. To date, few students have enrolled in the Masters course with a science background. Therefore only a handful of students are equipped on entry with a good understanding of molecular science and perhaps even fewer comprehend the complexities of the mechanical properties of materials. These demographics are not unique to the Dutch conservation course and must be a factor when designing course/ module content.

Experiential learning: teaching adhesive principles

All enrolled MA students at the University of Amsterdam are required to complete two science modules within their first year of studies. These modules run parallel with equivalent modules that include the topics of studio practice and cultural studies. Knowing and understanding the basic principles of polymer science is the first step along the 'road to knowledge' of adhesive technology and thus forms one of the key blocks within the first science module. The conservation student must have a sound understanding of molecular structures, valence forces of chemistry, intermolecular forces, thermodynamics, polymerisation and phase change. These core elements are taught by way of lectures, tutorials, revision and assessment and, while these topics are returned to later in the programme, it is expected that the basic knowledge will be fully grasped in this initial module.

Within this module theoretical knowledge is enhanced through experiential learning. This

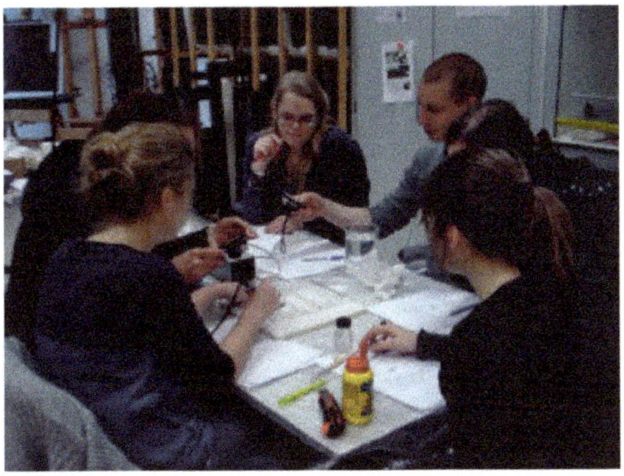

Figure 3 Group learning: stimulating cross-discipline interaction, creativity and divergent thinking.

allows the student to put into practice encountered theoretical concepts before taking this understanding into the studio. Experiments can be set up within a controlled environment that enables the student to visually grasp and assess addressed ideas. These interlinked experiments highlight aspects such as:

- film and bond formation
- adhesive strength: cohesion and adhesion
- handling and application properties: penetration, wetting properties, mobility, plasticity
- viscosity
- shear, peel, tensile and compression strengths.

A series of standard experiments is used that allows students to empirically assess the properties of selected adhesives (Duffy 1989).[5] The aim is to facilitate an understanding of adhesive characteristics and their suitability for different tasks. Subsequently, emphasis can be given to a systematic methodological approach, highlighting problem identification and problem solving, decision making, scientific inquiry and material testing related to individual projects. As it is often experiential learning that students recollect more readily, it is essential that these tests be designed for full impact.

In the laboratory students are encouraged to work in groups, stimulating cross-disciplinary interaction between the eight different specialties. Group work promotes creativity and divergent thinking, allowing students to realise that there may be more than one approach to a specific problem. Moreover, students from different backgrounds are able to compare the merits of individual adhesives that may not be commonly used within their specific discipline. Students are encouraged to critically evaluate published studies and alternative perspectives, and to test the consistency of their own experimental results against those reported in literature. Ultimately it is hoped that the students' outlook broadens, in order that they may anticipate future challenges and search more consciously for creative solutions to facilitate adhesive choice and trouble-shooting. Key didactic methods used include: 'think outside of the box' and 'thinking it through', promoting evaluation, critical thinking and reflection as well as stimulating all three learning domains (Bloom 1956).[6] Feedback and constructive criticism are essential elements of the workshop, which encourages an exchange of ideas. Practical sessions allow students to estimate the challenges of real conservation practice, gain a maximum awareness of their strengths, highlight any areas for improvement and identify actions that need to be taken to ameliorate performance. These aspects combine to provide a typical educational paradigm in which the importance of integrated teaching, learning and assessment has equal weight.

Experience of teaching students at MA level suggests that active participation in higher performance learning experiences, providing 'procedural knowledge', produces better learning outcomes, but these experiences must be supported by more passive activities in which 'declarative knowledge'[7] is given. Learning is at its most powerful when it is 'authentic' (Miller 1990). Practical tasks allow the student to move through the pyramid of competence, from being given knowledge (the novice), to being asked to remember and comprehend it (the advanced beginner), to understanding and applying that information (the competent). The following step is to encourage the students' ability to analyse and critically assess or reflect on results (the proficient), and ultimately develop the skills necessary to implement findings in real-life scenarios (the expert) (Bloom 1956; Dreyfus and Dreyfus 1980). It should be noted, however, that practical assessments, while they can offer high educational impact, might not be as reliable as other assessment forms, such as written papers or examinations. The student should therefore be evaluated and graded using both systems.

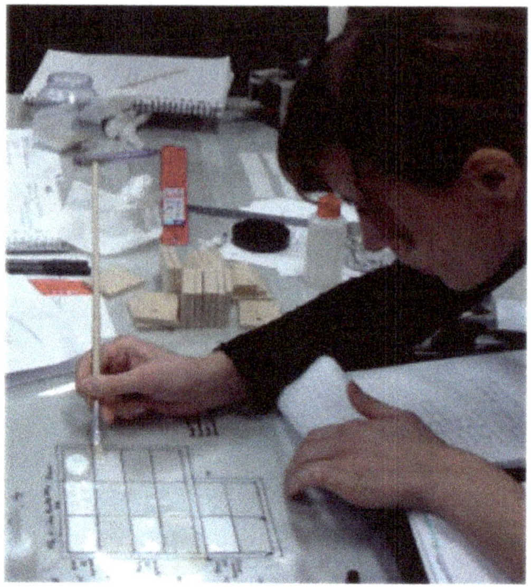

Figure 4 Adhesive application performance: studying the subtle differences in PVA adhesives.

Practical experiments

Simple tests can be used to simulate adhesive performance, giving students insight into why one adhesive may be selected over another for a particular task. Experiments can be repeated later with little expense or the requirement of intricate specialised equipment, hopefully encouraging the student to continue with these testing scenarios as part of his professional practice. Concepts learnt can easily be transferred to other adhesives with little additional effort. The properties of the polymers only are emphasised at this stage to reduce variables and allow the students to comprehend more fully adhesive performance.

Experiment 1: Application performance (PVA comparisons)

This experiment compares and explores the range of viscosity and tackiness of PVA-based adhesives in relation to molecular weight and additives. The degree of polymerisation and molecular weight, as well as the presence of any additives, will determine and affect many of the properties of the polymer including its cohesive and adhesive potential, its viscosity in solution, its elastic/ plastic potential, its solubility and its glass transition temperature (Bria 1986; Down 1996).

Students are required to apply three types of PVA dispersion adhesive (Evacon R, Mowilith DMC2 and a commercial PVA adhesive) to a non-absorbing surface.[8] While doing so they are encouraged to note the viscosities of the three adhesives, the ability of the adhesive to 'wet' the surface, the speed at which a film forms, its tackiness and the colour of the adhesive on application. The manner in which the film dries (in this case, solvent evaporation) and effect on volume (shrinkage) while drying are described. Once the films have formed each adhesive is reassessed in terms of tackiness, colour, transparency and success of application. Differences between the three adhesives are highlighted and discussed. Additionally, the students are asked to protect half of the dry film with a sheet of black paper and expose the adhesive films to direct sunlight for a number of months. While the results of this later element are impossible to discuss within the allocated time using the students' own examples, samples from previous years can be drawn upon.

The main goal of this experiment is to highlight how viscosity can be a factor of concentration, molecular weight, temperature of application and of substrate, as well as molecular structure. Adhesives are required to spread

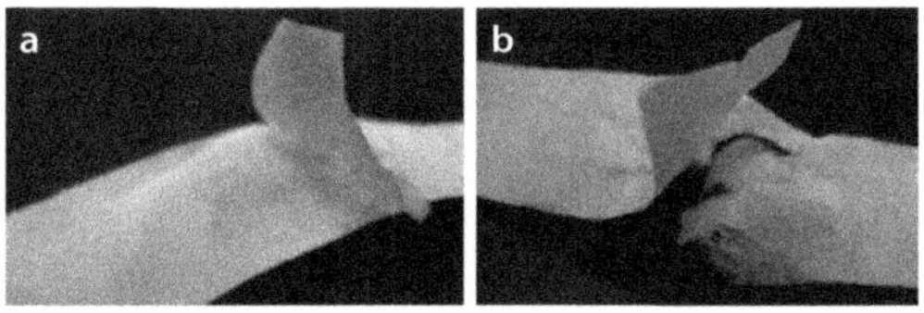

Figure 5 Bond formation and strength – hot melt versus solvent adhesives. Peel tests are performed to compare results between different adhesives: (a) beeswax and (b) Tylose C6000 20% solution in water.

out over a surface in order to facilitate bonding between two substrates. Thus an adhesive must have a low enough viscosity to flow over a surface. An understanding of the thermodynamic properties of both the polymer and substrate is therefore needed. While students may grasp the theoretical concepts governing these properties, their learning experience is enhanced when these can be visualised. Seeing the differences in viscosity evident between Evacon R and Mowilith DMC2 underpins the fact that viscosity and bond strength increase with molecular weight, and that wettability decreases. Throwing the commercial PVA adhesive into the mix allows for a discussion regarding additives to formulations, included to improve performance, and the potential detrimental consequences that these additives may have for an art object. The longer-term evaluation of these adhesives allows students to consider the notion of chemical stability.

Experiment 2: Bond formation and strength (solvent verses hot melt)

This experiment strives to show the student how bonds are formed. An adhesive holds two substrates together by forming a film (phase change from a liquid to a solid) and bridging the gap between them. Addressed here are the influence of molecular weight and concentration on bond strength, as well as the degree of penetration and length of time required for bond formation.

Two adhesive types that form films through different phase changes are used to demonstrate this aspect: hot-melt and solution adhesives. The bond formation of waxes beeswax and Cosmoloid H80 (both categorised as hot-melt adhesives) and Na-carboxy methyl cellulose Tylose C6000 at a 5% and 20% aqueous solution (classified as a solvent adhesive at different solid contents) are compared. The students are required to adhere two pieces of heavy ply Japanese paper together using these four adhesives, noting how each adhesive cures, how extensively it penetrates the adherends, and how a bond is formed. The activation temperatures of both waxes are recorded, as is the time required to achieve bond formation with all four adhesives. Once a bond has formed, the students reverse this bond using either a shear or peel action. The results are observed: does the paper tear or does the adhesive fail? Has the bond failed within the adhesive or at the interface? Pre-made examples are used for comparison with the students' own experiments, and to avoid confusion if their experiments fail.

This experiment has two objectives: first, time becomes a factor as cure time may be required to achieve a good bond and secondly, how penetration into the substrate affects bond performance. Bond formation relies heavily on mechanical interlocking occurring between the adhesive and the substrate, which can only happen if the surfaces of the two substrates are in close vicinity to each other and remain so while phase change occurs. To achieve good penetration, the adhesive must have the correct rheological properties, and must therefore be a mobile, low-viscosity liquid of low surface tension (Allen 1984). The experiment shows that, when applied at the correct activation temperature, the two hot-melt waxes flow into the paper substrate more successfully than the Tylose C6000 solutions. Comparison between the Tylose C6000 solutions reveals that the adhesive with the lower concentration flows further into the substrate but produces lower bond strengths. Results indicate that more damage occurs to the paper when reversing the hot-melt adhesives than when reversing the more viscose Tylose C6000 solution. Bond

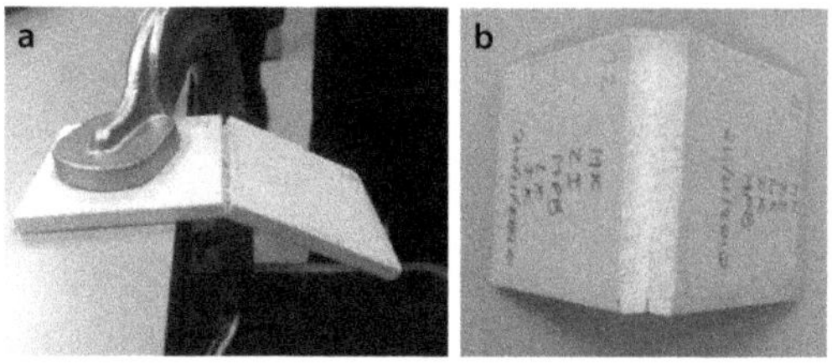

Figure 6 Mechanical properties: a joint performance/ bend test is performed to compare results between adhesives. (a) The load at which the bond fails is recorded and (b) the adhesive application on both sides of the joint is evaluated.

strength will increase if sufficient interlocking occurs, though these bonds are more difficult to reverse and doing so may cause damage to the substrate as a result of the physical force required to enable separation. It is conceivable that this factor will be paramount when selecting an adhesive for a particular task.

Experiment 3: Mechanical properties (joint performance)

Understanding the stresses, namely tension or compression, shear or cleavage, or peel, to which a joint may be subjected, is pivotal in adhesive selection. Choosing the right adhesive in terms of bond strength and rigidity/ flexibility is the crux of the matter. This experiment allows the student to apply stress to a pre-glued joint in order to evaluate the force required for that bond to fail and to compare data statistically between different adhesives tested. The experiment is set up using equipment readily found within a conservation studio, and emphasises empirical findings rather than the acquisition of scientific data.

Each group of students is given 10 quasi-identical, pre-broken samples of either wood or ceramic tile. The student is required to apply an adhesive to one side of the joint. This is applied to the whole surface area on five samples and to the central section (50% surface area) on the other five samples. Each group is given a different adhesive to provide a comprehensive set. Once a successful joint has been achieved the samples are subjected to tensile or bend tests by applying a load. The load at which each bond fails is noted and converted into newtons (SI unit of force) and pascals (SI unit of pressure).[9] Results are recorded and analysed. The data produced can be used to determine the force required to break the bond achieved by each individual adhesive. The disjoined sample is studied at high magnification to establish if the adhesive was applied successfully to the surface and whether any air bubbles exist within the adhesive film, and where the bond failed. Any lack of surface coverage is incorporated into the resulting calculations.

The purpose of this experiment is to allow the student to compare the bond strengths of various adhesives. More importantly, the student becomes aware that the strength of the adhesive is irrelevant to the area covered. Moreover, it is hoped that the student will fully comprehend the difference between stress and strain.[10] Choosing the right adhesive in terms of bond strength is crucial in conservation practice where reversibility and object integrity are paramount. It is essential that a full awareness exists of the elastic and plastic properties of polymers, their viscoelastic tendencies and potential for hysteresis, as well as their capacity to fail in either a ductile or brittle fracture. That these mechanical characteristics are both temperature and time dependent should not be forgotten and are highlighted within the experiment.

Conclusions

Evaluation of a range of parameters for the choice of adhesive for different purposes is a crucial learning experience that is subsequently implemented in daily practice throughout a conservator's working career. Designing a learning platform, which gives the student the required competencies to make informed choices, is critical when selecting course/ module content, especially given the time restraints imposed in an academic educational programme. Certain concepts can be taught through lectures, tutorials and by revision, however a more authentic experience may best be taught by experiential learning combined with appropriate feedback and evaluation. Through a series of experiments within a workshop environment the student is able to visually assess the concepts broached within the classroom. Within the Dutch conservation programme, these theoretical lectures and practical workshops are held early in the first semester in order that the student is provided with sufficient information before beginning treatment on conservation projects. Key teaching factors include feedback sessions to assess both group and individual results in comparison with those of well-established practitioners. Knowledge and understanding of taught concepts is deepened throughout the Masters and subsequent postgraduate phase.

Teaching adhesive theory to conservation students is of course broader than simply instructing them in the choice of one adhesive over another. It is a vast undertaking and the immense amount of information to be digested may overwhelm a student when these concepts are first encountered. This is especially true if the student arrives in the conservation studio directly from the art history lecture halls. The education of conservators has, over the last few decades, moved steadily towards a more professional and academic standing. Thus it is no longer appropriate to instruct solely novice conservators to use Glue A to carry out Task I or that Glue B fulfils the requirements to undertake Task II. Furthermore, as professionals we should no longer be limited to commercial products containing unknown ingredients, but should elect to use adhesives engineered and modified for specific use. Ultimately, students, as they become professional conservators, must be able to make those choices for themselves based on knowledge of the materials that they are using and the objects they are treating. The workshop environment aims to provide the student with the essential evaluative skills to implement these choices. The success of the practical sessions lies in the preparation of the experiments by the instructor, and the critical evaluation of the results by both the student and the instructor.

Lastly, the practical workshop allows the instructor and student to evaluate and assess the latter's capabilities and performance in a real-life context. Levels of competency and expertise should be judged in context and can be built upon, not only during subsequent training and modules, but throughout a conservator's career. It is implicit that on completing their educational requirements students are not experts within the field of conservation – only that they have gained sufficient information and skills to progress along a learning path.

> *Sine scientia ars nihil est* – without knowledge, skill is nothing

Acknowledgements

I would like to thank Bill Wei (RCE), Piet Iedema (UvA), Gwendoline R. Fife (SRAL), Norman Tennent (UvA), Nico Broers (St. Luc Academy, Liege), Ellen Jansen and Karolien Withofs (PG UvA 2010-2012) and the UvA Master 1 students: 2009–2011.

Notes

1. Source: Jean Mignot, French architect, 14th century,
2. Gustav Berger initiated a research programme to formulate an 'ideal' lining adhesive for paintings in 1967. The resultant hot-melt adhesive, Beva 371, has become one of the most predominant adhesives used within the conservation field. More recently in the 1980s Conservation by Design Limited developed a non-plasticised ethylene vinyl acetate dispersion adhesive, Evacon R, for laminating museum mount boards.

3. According to Caroline Keck (1977), 'the competent practitioner is always that person who selects the procedure most appropriate to the nature of the demand. "Appropriate" implies personal response to observed characteristics. "Select" implies freedom to make an educated choice.'
4. George L. Stout used the phrase a 'three-legged stool' to describe the interdisciplinary collaboration between science, art history and practice within the conservation field. Source: Joyce Hill-Stoner (Winterthur Program in Art Conservation, University of Delaware).
5. Examples of the following adhesive types are used: vinyl acetate and acrylic dispersions, cellulose ethers, waxes, cyanoacrylics and epoxy resins.
6. Bloom's learning domains are: Affective Learning, Cognitive Learning and Psychomotor Learning. These form a pedagogical framework for educational goals in which six categories (Knowledge, Comprehension, Application, Analysis, Synthesis and Evaluation) form a continuum from simple to complex, concrete to abstract, which a student completes in order to be able to acquire skills and abilities and put these into practice.
7. Declarative knowledge provides statements of fact, such as: Mowilith DMC2 is a polyvinyl acetate whose polymer consists of ethylene and vinyl acetate monomers. Procedural knowledge imparts the 'knowing how', such as: how to apply Mowilith DMC2 to achieve a good bond between two substrates.
8. Typically a sheet of Melinex is used.
9. A pascal is a measurement of force per unit area, defined as one newton per square metre.
10. Stress equals the force applied per unit area. Strain equals the magnitude of deformation occurring as a result of applied stress.

References

Allen, K.W. (1984) 'Adhesion and adhesives: some fundamentals', in *Adhesives and Consolidants: Contributions to the 1984 IIC Congress, Paris*. London: IIC, 5–12.

Berger, G. and Zeliger, H. (1984) 'The procedure of developing an adhesive for paintings: the importance of valid tests', in *Adhesives and Consolidants: Contributions to the 1984 IIC Congress, Paris*. London: IIC, 13–17.

Bloom, B.S. (1956) *Taxonomy of Educational Objectives, Handbook I: The Cognitive Domain*. New York: David McKay Co. Inc.

Bria, C. (1986) 'The history of the use of synthetic consolidants and lining adhesives', *WAAC Newsletter* 8(1): 7–11

Down, J.L., MacDonald, M.A., Tétreault, J. and Williams, R.S. (1996) 'Adhesive testing at the Canadian Conservation Institute: an evaluation of selected poly(vinyl acetate) and acrylic adhesives', *Studies in Conservation* 41(1): 19–44.

Dreyfus, S. and Dreyfus, H. (1980) *A Five-stage Model of the Mental Activities Involved in Directed Skill Acquisition*. University of California Berkeley, Operations Research Center Report, www.stormingmedia.us/15/1554/A155480.html (accessed 14 September 2012).

Duffy, M. (1989) 'A study of acrylic dispersions used in the treatment of paintings', *Journal of the American Institute for Conservation* 28(2): 67–77.

Horie, C.V. (2010) *Materials for Conservation: Organic Consolidants, Adhesives and Coatings*. Oxford: Butterworth-Heinemann.

Keck, C. (1977) 'Lining adhesives: their history, uses and abuses', *Journal of the American Institute for Conservation* 17(1): 45–52.

Miller, G. (1990) 'The assessment of clinical skills/performance', *Academic Medicine (Supplement)* 65: S63–7.

Wolbers, R. (2008) 'Short term mechanical properties of adhesives: solvent and plasticizer effects', *Colour and Conservation Cesmar* 7: 111–18.

Author's address

Kate Seymour, SRAL, Avenue Ceramique 224, 6221 KX Maastricht, The Netherlands (k.seymour@sral.nl)

Evaluation of funori stability in preparation methods

Ting-Fu Fan

Introduction

Funori is a traditional adhesive from Japan that has been used in paper, wall and painting conservation. Initially funori was only available and used as a conservation product in Japan. However, its usage has become more widespread, both as a result of intercultural projects, as well as the easy availability of various forms of funori through conservation manufacturers and suppliers worldwide.

What is funori?

The characters のり (*no-ri*) and 海苔 (*hai-ti*) both mean seaweed. Three species of seaweed are used to produce funori: *Gloiopeltis tenax* (Turner) Decaisne (1842), *Gloiopeltis furcata* (Postels & Ruprecht) J.Agardh (1851) and *Gloiopeltis complanata* (Harvey) Yamada (1932). In Japanese called *ma-funori*, *fukuro-funori* and *hana-funori* (Mitsuo 2002), each one has a characteristic appearance and size (Figure 1). These three seaweeds are considered a subset of the family Endocladiaceae, which belong to the order Gigartinales, under the class of Florideophyceae. According to investigations by marine biologists, these seaweeds can be found on the Japanese coast, as well as North America, Chile, China, Taiwan and Vietnam.[1]

Chemical composition of funori

Funori is a sulphated polysaccharide extracted from the seaweed composed of repeating units of the monosaccharides β-D-galactose-6-sulphate and 3,6-anhydro-α-L-galactose or 3,6-anhydro-α-D-galactose. Funori has been classified by its chemical structure as an agar type of funoran, but funori from *G. furcata* and *G. complanata* contains both agaroid and carrageenan chemical structures (Takano *et al.* 1995, 1998). The main chain of polysaccharides are linked by β (1→3) links and because of its highly sulphated nature this makes the solution different to other low sulphated polysaccharides such as agar, which become solid at room temperature. This is also the reason why funori can be used as an adhesive and consolidant in conservation (Horie 1987; Swider and Smith 2005).

Funori as an adhesive

The funori sheets used in traditional eastern conservation were produced in Japan. The funori sheets also called 板ふのり (*ita-funori*) are made from two of the three abovementioned seaweeds: *G. tenax* and *G. furcata*. The manufacture procedure includes collecting, washing, bleaching, and exposing the seaweeds to the sun to dry and bleach (Nariaki 1999). In practical applications a hot extraction method

	Scientific name	Japanese		Chinese	
1	*Gloiopeltis tenax* The type species (holotype) of the genus	ma-funori	真布海苔	hai-luo	海蘿
2	*Gloiopeltis furcata*	fukuro-funori	袋布海苔	lu-jiao-hai-luo	鹿角海蘿
3	*Gloiopeltis complanata*	hana-funori	花布海苔	–	–

Figure 1 The nomenclature of funori.

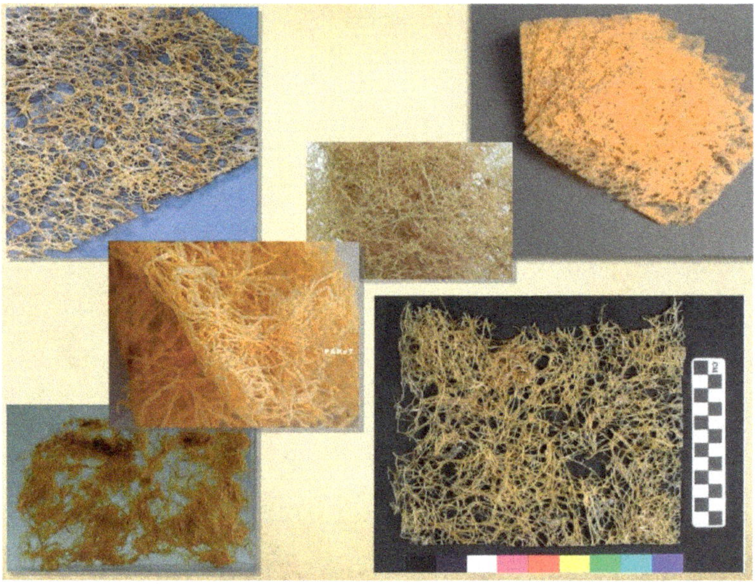

Figure 2 Funori sheet products.

is usually used. Traditionally funori sheets are placed in water and boiled for a period of time, stirring when necessary. Impurities are then filtered out to make a glutinous adhesive solution. Cold extraction can also be used to produce a weaker funori adhesive. The quality of traditional funori sheets can be distinguished mainly by appearance: the colour of the product, the size and completeness of the seaweed, and its cleanliness (Figure 2). However, as it is a natural product there can be variations in quality. Research has found that even seaweeds grown in the same area often differ in composition from year to year. The seaweed can be affected by variations in the seasonal conditions (Ji 1997). Furthermore, the quality of the seaweed acquired can vary between different factories in different production areas; variations in manufacturing procedures can also affect the quality of funori sheet products.

To minimise variation in quality, JunFunori, a standardised funori product, was developed by conservators and scientists in the laboratories of EMPA (Swiss Federal Laboratories for Materials Science & Technology) in collaboration with Lascaux (Michel *et al.* 2002; Geiger and Michel 2005).

Application in painting conservation

In eastern painting conservation, funori has been used not only to locally consolidate

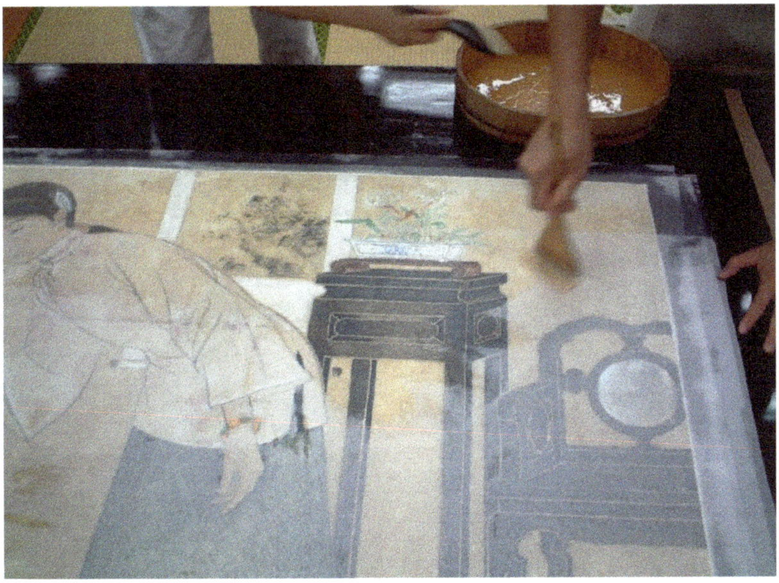

Figure 3 Applying funori as a facing.

pigment and painting layers, but also on large areas or even entire painted surfaces as a temporary facing adhesive, in particular to protect the surface of silk paintings during backing removal (Figure 3). With the exception of eastern paintings, funori has also been used in paper conservation projects. However, paste has been used extensively for a number of centuries as a temporary facing adhesive in China. To protect the paint surface during the removal of backing layers, paste is applied to adhere facing materials to the front of the painting to provide temporary protection. This facing is then removed after conservation has been completed. Many eastern painting conservators do not consider it necessary to completely clear the funori adhesive from the surface of a painting after the facing has been removed, instead choosing to remove only visible excess adhesive that has an effect on the surface appearance. However other conservators remove as much adhesive as possible in order to reduce any negative effects that the residual adhesive might have on the painting surface. There has not been a formal study to determine whether it is necessary to remove funori when used as a temporary facing adhesive on traditional eastern paintings, but from practical experience, conservators generally agree that as much adhesive as possible should be removed from the paint surface (Fan 2004).

Although eastern and western painting materials and styles differ, the factors and the reasons for selecting adhesives for consolidation and facing are very similar. For example, properties such as low gloss and low contractibility are often desirable. As such, funori has been used for the conservation of objects, maps, matt paintings and other artworks.

The application of funori

There are three main aspects to consider when using funori: how to prepare it, how to ensure its stability, and how to retain its purity. There is no standard process for preparing funori as a conservation adhesive. However, many tests have been carried out to find the most effective preparation technique, some of which have been published. One of the most

Table 1 Traditional funori sheet preparation methods.

Adhesive	Dry funori	Water	Rinse	Soak	Change water	Heat	Filter	Precipitate
F1	2 g	50 ml	–	–	–	80°C for 1 hour, stir occasionally	Strained twice	
F2	2 g	50 ml	Under dripping distilled water for 30 minutes	–	50 ml	80°C for 1 hour, stir occasionally	Strained twice	
F3	2 g	50 ml	Under dripping distilled water for 1 hour	8 hours	50 ml	80°C for 1 hour, stir occasionally	Strained twice	With 4 volumes of pure ethanol

Table 2 Other adhesive preparation methods.

Adhesive	Material	Water	Soaked and dissolved	Treatment	Rest	Storage
JunFunori (EMPA)	2 g	50 ml	In warm water for 2 hours	Stirred occasionally	Overnight	5 vol% isopropyl alcohol
Carrageenan (Jacquard)	2 g	50 ml	In warm water for 2 hours	Stirred occasionally	Overnight	–
Methyl cellulose	2 g	50 ml	In 50 ml water for 2 hours	Stirred occasionally	Overnight	–

effective methods of preparing funori is based on traditional eastern studio practice. Thin, dried sheets of prepared funori adhesive, the colour of light amber, are pre-prepared and stored until required for use, when they can be dissolved either at room temperature or immersed in hot water before application. Significant research has been carried out on successful methods of funori preparation by the Freer Gallery of Art in Washington DC (Swider and Smith 2005).

JunFunori, a purified funori product, was developed in order to avoid the salt and acid/ alkali problems caused by impurities in traditional funori sheets. JunFunori is considerably more expensive than traditional funori sheets but it provides good stability under ultraviolet-accelerated ageing tests and is an effective consolidant for matt paintings and objects. Due to the considerable amount of funori used in some conservation treatments, conservators generally still tend to use traditional funori sheets, rather than the high-priced JunFunori.

In practice funori is usually prepared in small amounts every few days or when needed, rather than storing larger volumes for long periods of time, and the method of producing the funori adhesive is kept consistent throughout a particular treatment process.

Scientific testing

This experiment will be separated into several parts. First, based on case study papers and practical experience, the different methods of preparing funori will be discussed, alongside the preparation methods of JunFunori and carrageenan, another form of seaweed-derived adhesive, to compare the results, and find the most effective preparation method. Methyl cellulose was used as the control throughout the experiment.

Test setting: adhesive preparation

Traditional funori sheets were prepared using three different procedures. Preparation methods are listed in Table 1 and other adhesives in Table 2. All six adhesives were prepared in solutions of 4 g adhesive/ 100 ml water, based on practical facing adhesive requirements.[2]

The prepared adhesives were kept in transparent measuring bottles to observe the colours of the adhesive solutions. A fixed amount of each adhesive was applied in both thick and thin layers on separate glass slides. White sheets of paper with black lines and characters on them were placed under the glass slides so that changes to the adhesive colour and transparency could be more easily distinguished.

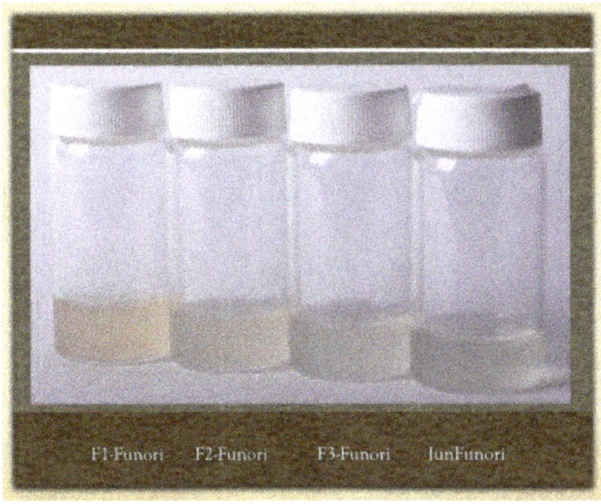

Figure 4 Different funori solutions, from left to right: F1-Funori, F2-Funori, F3-Funori, F1-JunFunori.

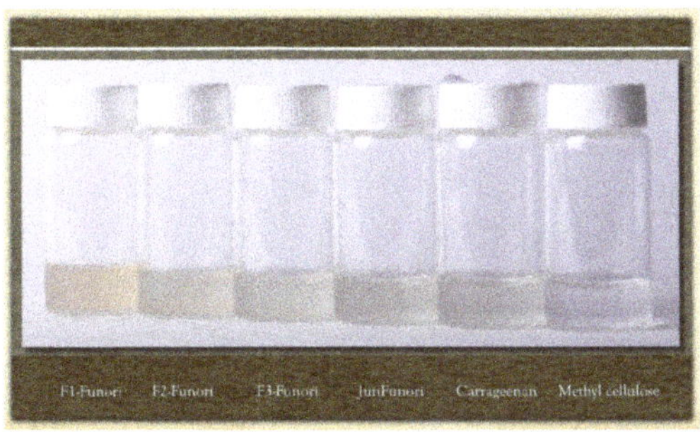

Figure 5 All adhesive solutions from left to right: F1-Funori, F2-Funori, F3-Funori, F1-JunFunori, carrageenan, methyl cellulose.

Accelerated ageing

Accelerated ageing tests were used to simulate conditions of materials after ageing in a relatively short period. It was hoped that this would provide some evidence of how the materials would change over time, and analyse potential problems that may arise. All of the sample types were subjected to two separate forms of accelerated ageing: ultraviolet (UV) and relative humidity (RH) and heat. Colour changes were also measured.[3]

The period of UV exposure was 12 days (288 hours) and 22 days (528 hours). During the experiment all adhesives were facing the light source. The temperature was 38°C at 48% RH. The humidity- and temperature-accelerated ageing was at 85°C at 87% RH for three days (72 hours), six days (144 hours) and 12 days (288 hours).

Results and discussions

Comparison of adhesive solutions

The aim of the experiment was to compare the differences in colour of the differently pre-prepared adhesives and attempt to estimate the influence that these would have on the appearance of a painted surface. JunFunori, when compared to the three other pre-prepared traditional funori solutions, was found to be more transparent with a light amber colour, showing the significance of the purification technique. F1-Funori solution, which was heated directly without any pre-processing, was found to have the darkest colour of all the types of funori tested. F3-Funori solution with the most extensive

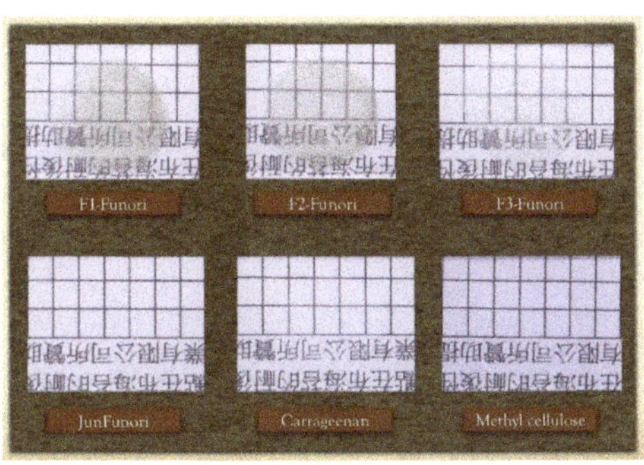

	Δ*(CIE 1976)			
	ΔL*	Δa*	Δb*	ΔE*
F1-Funori	-6.60	-2.20	12.10	13.96
F2-Funori	-6.44	-1.99	8.22	10.63
F3-Funori	-6.87	-1.07	4.22	8.13
JunFunori	-6.04	-0.76	3.15	6.85
Carrageenan	-6.23	-0.39	2.12	6.59
Methyl cellulose	-1.10	0.07	1.72	2.04

Figure 6 Test samples and colour difference.

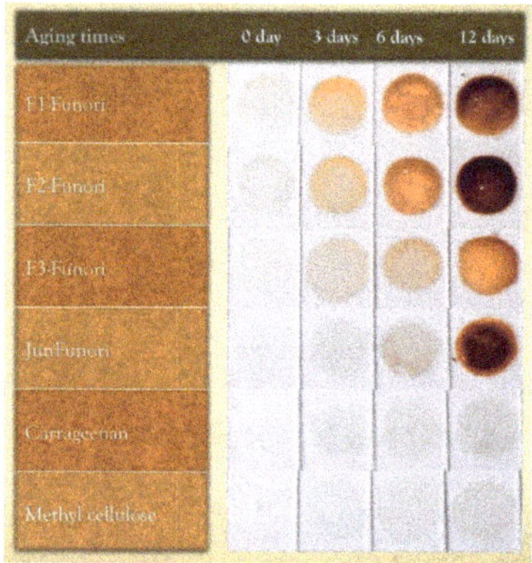

	ΔE^*			
	0 Days	3 Days	6 Days	12 Days
F1-Funori	13.96	22.60	69.52	63.73
F2-Funori	10.63	23.60	61.52	68.36
F3-Funori	8.13	10.93	22.46	61.56
JunFunori	6.85	4.99	16.29	68.26
Carrageenan	6.59	2.32	2.36	3.20
Methyl cellulose	2.04	4.02	4.39	5.37

Figure 7 Humidity and heat-accelerated ageing samples and colour difference.

preparation process showed good results and was only slightly more yellow than JunFunori (Figure 4). Although JunFunori produced the best results, traditional funori sheets produced a light coloured and clean adhesive solution if they were sufficiently pre-prepared.

Comparisons were also made between Junfunori, carrageenan and methyl cellulose adhesive solutions. Methyl cellulose solution was found to be the most transparent and although Junfunori and carrageenan are only slightly different, the latter solution had a more transparent appearance (Figure 5). However it should be noted that when the solutions were made up at 2% concentration almost no difference could be seen between the three solutions.

Transparency tests

To gauge the differences in appearance of the adhesives, a visual simulation test was conducted. Each adhesive was applied in both thick and thin layers on separate glass slides. The slides with a thick layer of adhesive were placed on a white board, which was printed with black lines so that the adhesives could be examined over light and dark surfaces.

Adhesive transparency and colour was observed from different angles and images were taken from a vertical angle so that the differences could be better observed. From the test, methyl cellulose was found to have the least optical interference, being almost transparent with the printed lines underneath clear to see. JunFunori showed the second best penetration with only a very slight yellowish tinge, but remained transparent with a more matt appearance. Although carageenan was found to be less transparent and slightly cloudy (like looking through a light fog), the printed lines were still clearly visible. In comparison the F1-Funori sample without any pre-preparation was the least satisfactory in terms of transparency and colour change. However the pre-prepared F3-Funori sample gave an acceptable result (Figure 6). The same observations were made with the thin layer adhesive samples. Since there was a very small amount of adhesive and the layers were very thin, each sample appeared almost transparent and colourless. Any difference and influence on the appearance of the pattern underneath the samples was barely noticeable.

Humidity and heat-accelerated ageing

Although the reactions of the accelerated ageing were extreme for some samples – probably as a result of the high temperatures used – the trends observed in these changes are still informative. Carrageenan and methyl cellulose had only changed slightly after accelerated ageing (Figure 7).

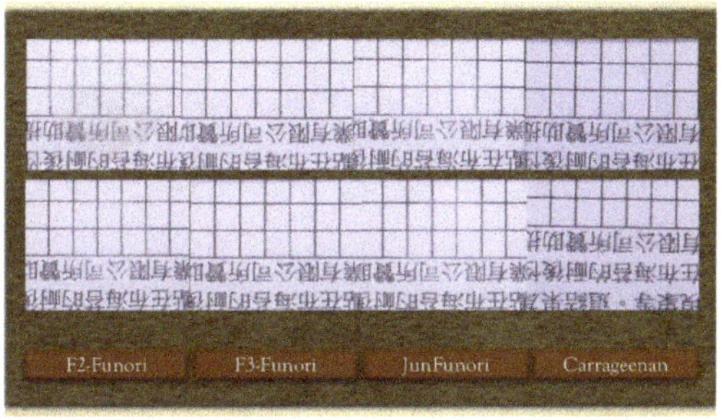

ΔE^*				
	0 Day/0 Hours	13 Day/312 Hours	16Day/384 Hours	22Day/528 Hours
F1-Funori				
F2-Funori	10.63	3.26	2.23	2.16
F3-Funori	8.13	1.98	3.17	1.29
JunFunori	6.85	1.26	2.73	1.89
Carrageenan	6.59	0.44	0.69	1.95
Methyl cellulose				

Figure 8 Ultraviolet-accelerated ageing samples and colour difference.

The result for the funori samples clearly showed that that the more pre-preparation or purification process to which the sample was subjected, the less colour change was noticeable in the sample. For example, F3-Funori and JunFunori were less discoloured than F1-Funori, which had changed dramatically in colour during the first stage of the test, turning into a dark brown colour thereafter. On the last day of the ageing test, every funori-type adhesive had been affected in terms of colour change, however throughout the process of testing JunFunori was the least changed of all the samples. Carageenan was also found to be a very stable adhesive with a colour change similar to methyl cellulose, both of them showing better stability than traditional funori and JunFunori. Currently carageenan has not been used as an adhesive in conservation, however its similarity to funori and its lack of discoloration during testing indicate that further tests and experiments should be carried out to find a use for it within conservation.

The purpose of this test was not to describe the poor quality of funori or JunFunori, but more importantly, to understand and indicate the importance of pre-preparation.

Ultraviolet-accelerated ageing

The result for the samples subjected to UV light ageing did not replicate the results of the humidity and heat ageing. After the UV-accelerated ageing test, none of the originally slightly yellowed samples had darkened in colour – instead they became somewhat more transparent and colourless than they were before UV light ageing. A possible explanation could be that the UV light bleached some of the coloured materials in the funori (Figure 8).

Conclusions

Before ageing

- Traditional funori preparation methods are successful in filtering out the coloration/impurities.
- The more pre-preparation, the clearer the funori solution becomes.
- Among four kinds of funori solutions, JunFunori, which has undergone a more thorough purification procedure, showed its advantage through reduced colour in solution and dried film form.
- The colour of the carrageenan solution and its influence on colour appearance was found to be similar to JunFunori. However methyl cellulose had the least colour and most transparency in liquid and dried film form.

After ageing

- After two types of accelerated ageing experiments, the samples subjected to humidity and heat showed more discoloration than the group subjected to UV radiation. The adhesive samples in the

humidity- and heat-accelerated ageing group turned reddish brown, while the group that underwent UV radiation turned slightly misty (blanched). However this dramatic colour change seen in the humidity and heat ageing has not been observed in other experiments using artificial ageing and may be a reaction to the high temperatures used.
- Humidity- and heat-accelerated ageing experiments indicate the advantages of purification.
- Carrageenan samples showed little colour change in testing. There is no published literature on this material and more tests and research on carrageenan may prove to be worthwhile.

Acknowledgements

Many people provided advice and help during this research. This paper was one stage of a three-year research project that began in 2010, sponsored and supported by San Jian Art Conservation Co., Ltd., Taipei, Taiwan.

Notes

1. Further information on seaweeds, including scientific names, class and ecology can be found in the Algaebase database: www.algaebase.org/. See also www.funori.com.
2. Adhesives of 4% concentration are usually the highest used for silk paintings. However there is no standard concentration for conservators, who prepare their solutions on an individual case basis.
3. Colorimeter: UltraScan XE, HunterLab Inc., light source: D65, CIE 10 degree observer (1964).

References

Fan, T. (2004) *Facing Application: A Study of Temporary Materials*, unpublished Master's thesis, Tainan National University of the Arts, Taiwan.

Geiger, T. and Michel, F. (2005) 'Studies on the polysaccharide JunFunori used to consolidate matt paint', *Studies in Conservation* 50(3): 193–204.

Horie, C.V. (1987) *Materials for Conservation: Organic Consolidants, Adhesives and Coatings*. Oxford: Butterworth-Heinemann.

Ji, M. (1997) *Seaweed Chemistry*. Beijing: Science Press.

Michel, F., Geiger T. and Teoh-Sapkota, G. (2002) 'Funori, ein japanisches Festigungmittel für matte Malerei', *Zeitschrift für Kunsttechnologie ZKK* 16: 257–75.

Mitsuo, C. (2002) 千原 光雄, 日本の海藻：フィールドベスト図鑑. 日本, 学習研究社, 113 [*Seaweeds of Japan: Field Best Encyclopedia*. Japan: Gakken, 113].

Nariaki, Y. (1999) '修理に使用するる布海苔現狀の試作', 国宝修理裝潢師聯盟定期研修會講演集, 平成11年度: 12–9.

Swider, J.R. and Smith, M. (2005) 'Funori: overview of a 300-year-old consolidant', *Journal of the American Institute for Conservation* 44(2): 117–26.

Takano, R., Hayashi, K., Hara, S. and Hirase, S. (1995) 'Funoran from the red seaweed, *Gloiopeltis complanata*: polysaccharides with sulphated agarose structure and their precursor structure', *Carbohydrate Polymers* 27(4): 305–11.

Takano, R., Iwane-Sakata, I., Hayashi, K., Hara, S. and Hirase, S. (1998) 'Concurrence of agaroid and carrageenan chains in funoran from the red seaweed *Gloiopeltis furcata* Post. et Ruprecht (Cryptonemiales, Rhodophyta)', *Carbohydrate Polymers* 35(1–2): 81–7.

Author's address

Ting-Fu Fan, 11F, No. 100, Ligong St., Beitou Dist., Taipei 112, Taiwan (sjconservation.co@gmail.com)

Conservation of objects for the 'Arts of Thailand' display at the V&A: case study examining the complex treatment of two panel paintings depicting scenes from the last ten of the Buddha's former lives or Jatakas (Thai Thotsachat)

Sophie Reddington and Mike Wheeler

Introduction

In 2008 a new display was created at the Victoria and Albert Museum (V&A) in London funded by the Royal Thai government to celebrate the 80th birthday of His Majesty King Bhumibol Adulyadei of Thailand. A variety of significant objects from the V&A collection was chosen for the display among which were sculptures, jewellery and decorative pieces, as well as two panel paintings (although only one painting would be on display at any one time). These panels, created by an unknown artist in the early 19th century, were painted in gouache on wooden boards and depict scenes from the last ten of the Buddha's former lives or Jatakas. Both required treatment before going on display (Figure 1).

Condition

The panels were in very poor condition with dramatically lifting paint in many areas, and were therefore kept horizontal in order to avoid further paint loss. The paint itself also appeared to be lacking in binding medium, including areas of flaking where the layers of paint were lifting away from both the ground layer and from the wooden panel itself. In a few areas the paint was severely broken up. This necessitated remedial action to humidify the flakes in order to allow them to be pushed back down into their correct alignment prior to introducing a consolidant.

Support

The panels were made from tropical hardwood composed of single vertical boards, which varied in thickness throughout (Figure 2). Thin wooden mouldings were nailed to the outer edges of the panels to create frames for the paintings and these were painted red over a white ground. The mouldings had started to detach in places because both panels had a slightly convex warp.

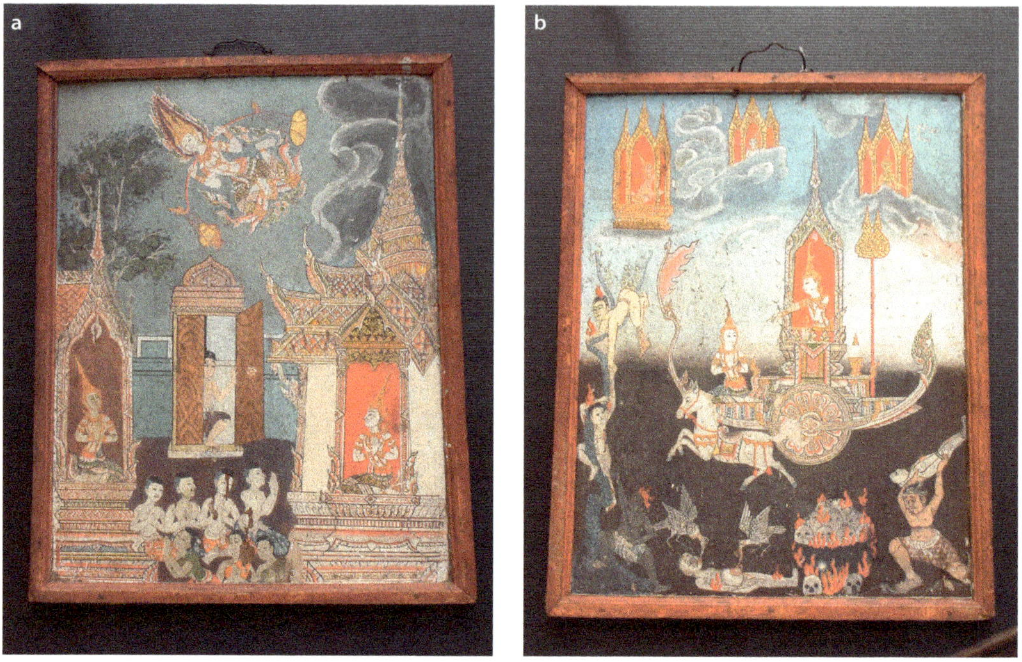

Figure 1 (a) IS.38-2005 Unknown Artist, Thai School, scene with figure in the sky: painting before treatment; (b) IS.39-2005 Unknown Artist, Thai School, scene with figure in chariot: painting before treatment.

Ground and paint layers

The thin, white, chalky ground appeared to be quite crumbly and dry. The paint was very fragile, brittle and flaking badly throughout. In some areas flakes of paint were raised in peaks and the wood was visible underneath (Figure 3). As is normal for gouache paintings there was no varnish layer, giving a matt surface appearance.

There was a considerable amount of overpaint and retouching on both paintings, particularly in the sky, which indicated that the flaking paint was a recurring problem. Areas of paint loss had also only been toned in without being filled.

Figure 2 IS.38-2005: reverse of the painting.

Conservation treatment

None of the usual methods of consolidation, such as the use of sturgeon glue and a heated spatula to reattach paint, could be used as

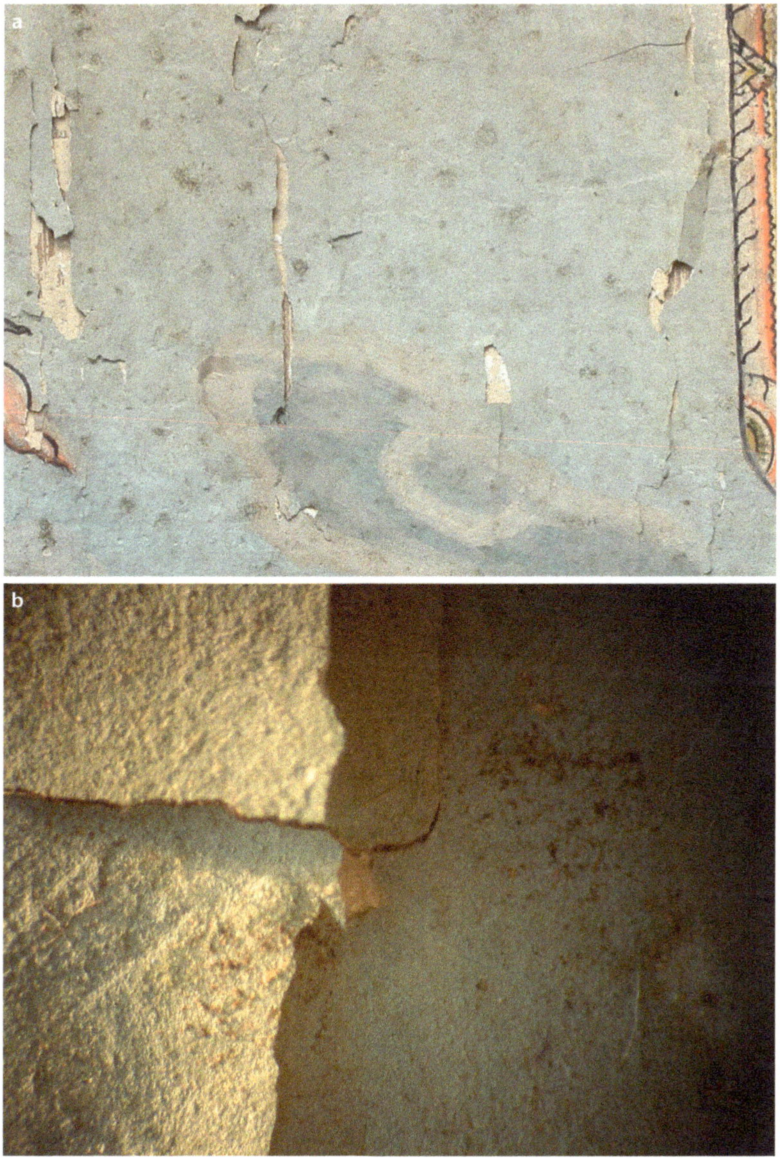

Figure 3 (a) IS.39-2005: detail of lifting paint; (b) IS.39-2009: microscopic detail of lifting paint.

any kind of pressure or moisture would have crushed the raised paint. Additionally an adhesive with a low viscosity could potentially have mobilised the loose paint. At the same time an adhesive was needed that would not stain the matt paint surface or leave it glossy.

The use of a nebuliser (Figure 4) to create a fine mist of adhesive for treating friable paint surfaces was developed by Sandra Grantham (1999).[1] She tested a number of traditional Japanese consolidants: nikawa and funori alongside cellulose ethers, ethyl hydroxy ethyl cellulose (EHEC) and low viscosity methyl cellulose. Measurements of the changes in refractive index brought about by the different consolidants led Grantham to conclude that

EHEC was the best choice for overall application to predominantly matt paint films. Nikawa and funori caused considerable changes to both surface gloss and to the refractive indices of the films to which they were applied. The advantage of the nebuliser over other methods of delivery used for aqueous consolidants is the relatively small size of mist droplets compared with an ultrasonic humidifier or a spray gun (Michalski and Dignard 1997). This is particularly relevant in the context of an overall application of consolidant to a matt paint film, as it is important to achieve a layer of consolidant of consistent thickness to avoid an uneven result. The smaller-sized droplets would also be more easily absorbed by the matt paint layer without leaving any water marks or stains. The nebuliser was originally developed for medical use to propel viscous liquids (e.g. antibiotics and other medicines) and is easy to use. Nebulisers are widely available for medical uses and are relatively cheap to purchase (approx. £80–£150). The model used at the V&A is a Pari Turboboy, which has been modified by fitting it with a side stream attachment. This model was chosen by Grantham for its robustness, portability and because the compressor is also very quiet in operation.

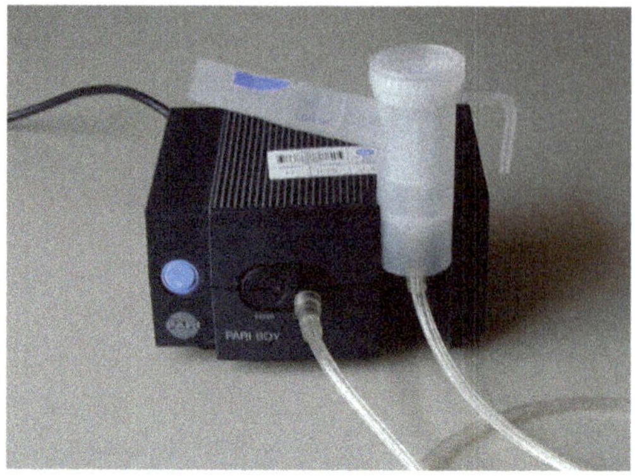

Figure 4 The nebuliser.

Consolidation

The consolidation of paint was undertaken in three different stages. EHEC was selected in preference to methyl cellulose, as although it has slightly lower resistance to yellowing, it is available in a much wider variety of grades, allowing a very low viscosity solution to be propelled easily in droplet form. Methyl cellulose, for example, becomes too weak an adhesive to be effective when diluted to the point where it can be easily nebulised. Extensive testing of samples has shown that a 0.25% solution of EHEC in deionised water is sufficiently strong provided that multiple applications are made, and that care is taken to overlap each pass of consolidant mist.

As a first step all of the paint was carefully treated with a 2% Bermocoll E230FQ solution in deionised water. This enabled the paint to soften slightly and secure it very slowly at the same time. A small amount of industrial methylated spirit (approx. 5% to the overall solution) was added to the consolidant to assist with penetration and reduce the likelihood of staining caused by pooling. Care was taken to limit the amount of alcohol added as this has been found to adversely affect the flexibility of the consolidated paint.

After the solution had been gently introduced into the paint in a very fine mist with a nebuliser (Figure 5), each area was then left briefly under the pressure of tiny weights with Bondina (non-woven, chemically inert 100% polyester release material) as a release sheet. Several areas of the paint were exceptionally distorted however, and required considerable careful manipulation prior to consolidation in order to align them correctly. In these instances the paint was locally humidified with the nebuliser containing only water. Magnification was required to position the fragments accurately. The flakes were then pushed into place with a sharpened wooden swab stick or the tip of a fine brush before applying light pressure to the affected area with fingertips used through silicon release paper.

After the success of the initial nebuliser treatment it was necessary to fill in the voids left in areas of paint loss with repeated applica-

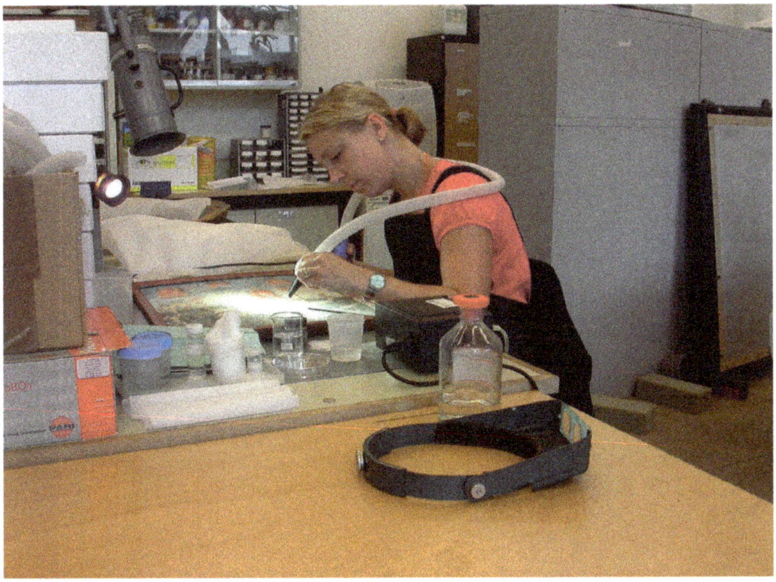

Figure 5 Working with the nebuliser.

tions of a stronger consolidant applied with a fine 00 paint brush. This stronger solution of 0.5% Bermocoll E230 FQ in deionised water was dripped into the paint voids to act as a filler with the aim of preventing further losses occurring from the exposed edges. In the next step the same (0.5%) Bermocoll solution was applied with a fine brush and fed underneath the lifting fragments of paint. This enabled the tenting paint to be gently pushed back into place.

While this method made it possible to manipulate and secure the extremely fragile paint it was decided that a stronger adhesive would be needed to ensure the long-term stability of the paint. Lascaux Medium for Consolidation 4176 (an aqueous dispersion of an acrylic polymer) was chosen for its very low viscosity, which allowed it to flow easily beneath the paint layers.

Examination of consolidated and unconsolidated areas revealed that even under magnification, it was not possible to detect any glossiness. On a paper substrate, the best results were achieved by overall humidification before treatment, and by slowing down the drying of the treated paint film to allow it to absorb as much consolidant as possible by capillary action. When applying this methodology to the Thai paintings however, the authors decided prior humidification of the panel might be risky and instead opted to cover the consolidated areas of the painting with polythene to slow down drying as much as possible.

The entire treatment took many hours of patient work to stabilise the panel and render it fit for handling and display. The matt appearance of the paint was maintained satisfactorily, and it was impossible to detect the presence of the consolidant even when examined closely under the microscope (Figure 6).

Cleaning

Having secured the paint, the surface dirt had to be removed. As the paint surface was too vulnerable to be cleaned with a damp swab, a special cleaning sponge called WL Slurpex (a PVA chamois block cleaning sponge) was used

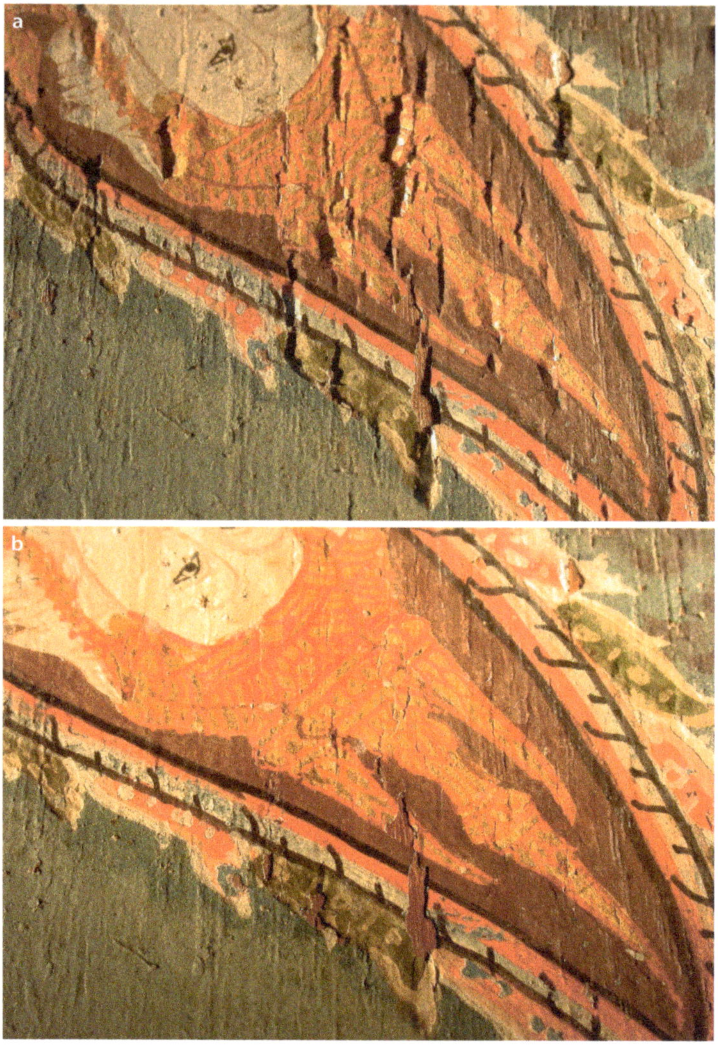

Figure 6 (a) IS.38-2005: detail of the figure in the sky photographed in raking light before treatment; (b) the same detail photographed after treatment.

to reduce the surface dirt. It could absorb the loose dirt without wetting the paint surface.

Retouching

Finally a few areas, where the white ground was exposed and old retouchings had badly discoloured, were retouched in watercolour (Figure 7).

Conclusions

Although the use of a nebuliser is not well known in paintings conservation, for this specific problem it proved to be the perfect answer. Using different materials and methods for different stages was beneficial rather than attempting to solve the problem all at once; each stage aimed to solve a different part of the problem. Analysing the different elements of the problem was a very important part of

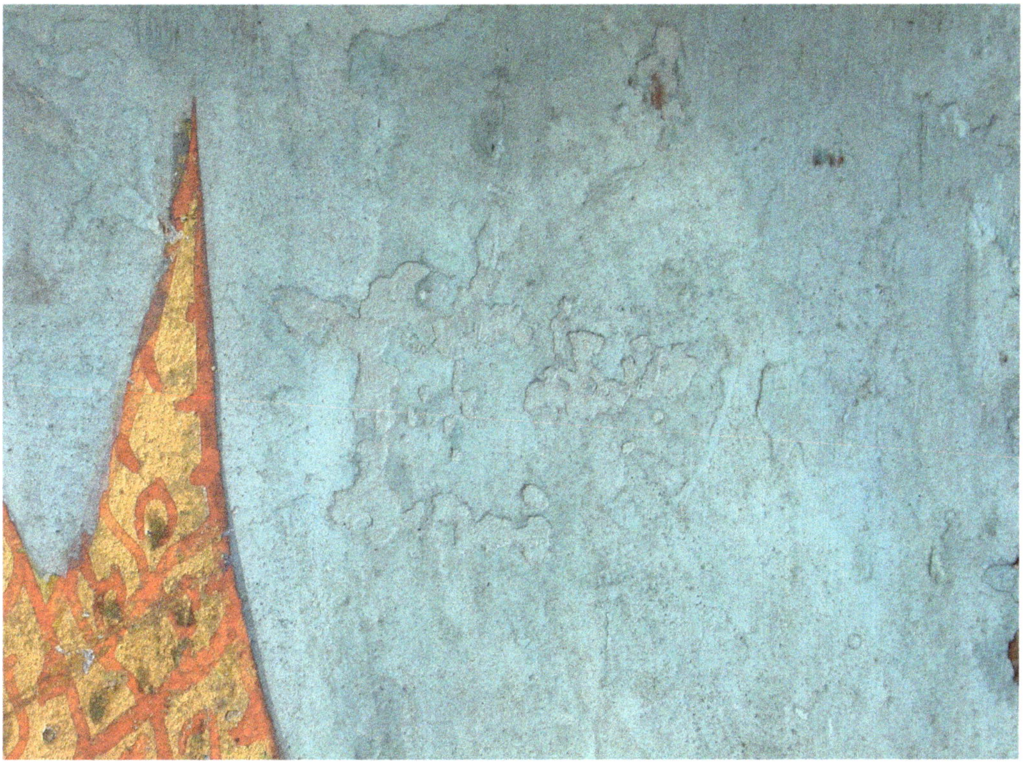

Figure 7 Detail in the sky after retouching.

the process. Furthermore it was interesting to be confronted with all the different types of cellulose ethers that conservators use in paper and book conservation. The project shows how useful and important it is to collaborate with conservators from other disciplines.

Note

1. During research carried out for her PhD in Paper Conservation in 1999; the first PhD awarded by the RCA/V&A course.

References

Grantham, S. (1999) *Byo-bu and Fusuma: Developing an Approach to the Conservation of Japanese Screens through Historical and Technical Study and an Investigation of Current Practices*, PhD thesis, RCA/V&A Conservation.

Michalski, S. and Dignard, C. (1997) 'Ultrasonic misting. Part 1. Experiments on appearance, change and improvement in bonding', *Journal of the American Institute for Conservation* 36: 109–26.

Authors' addresses

- Sophie Reddington, 21 Elms Road, London SW4 9 ER, UK (sophiereddington@yahoo.co.uk)
- Mike Wheeler, Conservation Department, Victoria and Albert Museum, London, UK

Facing adhesives for size-tempera painted wood: results of a research internship for Historic Scotland

Chantal Thuer

Introduction

The aim of the research project was to test the suitability of a range of adhesives for facing size-tempera painted wooden ceilings from the Scottish Renaissance (1550–1650). It is believed that the method of painting these ceilings was to apply a ground layer of chalk and size on the bare wood and then execute the design in subsequent layers of pigment mixed into size, creating a colourful yet matt appearance. However the precise nature of this size medium is unknown. The ceilings are generally located in buildings with relatively unstable environmental conditions therefore they are often subject to delamination of the paint layer from the wood (Figure 1).

Historic Scotland has a number of painted ceilings in its care, both in storage and *in situ*, as well as a sizable salvage collection, consisting mainly of open beam-and-board and barrel-vaulted construction. As part of their previous

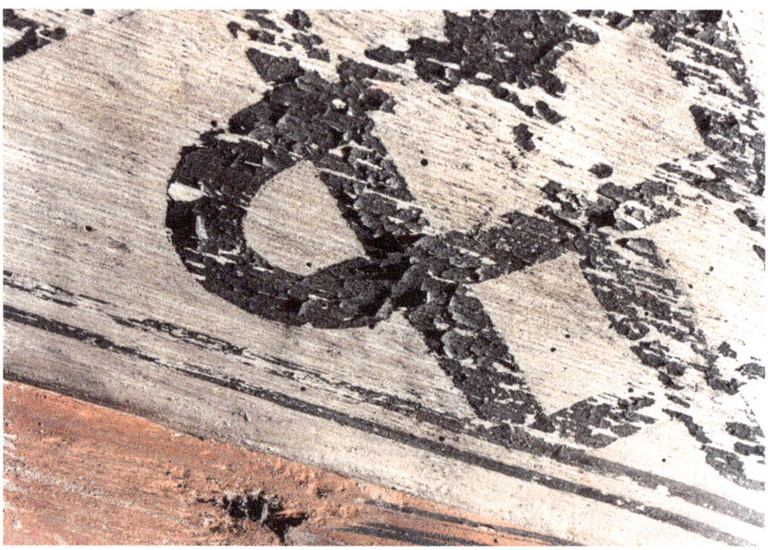

Figure 1 Flaking paint on the painted ceiling in the Huntingtower prior to consolidation. (Source: Historic Scotland, Crown Copyright.)

conservation treatment many of these have been consolidated or protected with a facing layer. The facings normally consisted of a layer of Eltoline tissue, with gelatine as the adhesive.

When applied to matt paint, a facing adhesive should be considered as belonging in the same category as a consolidant – that is, the material used will remain in the paint structure with the possibility of altering or damaging the paint layer. A facing adhesive applied onto matt absorbent paint can never be entirely removed, thus restricting adhesive options in further treatments. There is also a risk that an adhesive will alter the surface appearance of size-tempera media.

The aim of the internship was to test adhesives and facing materials to determine which materials might meet conservation standards. The focus of the research was to identify suitable facing adhesives for the most commonly encountered issues involved in the care of size-tempera ceilings, including transport, emergency measures and for general interventive treatments.

Selection of adhesives

A review of conservation literature was carried out at the beginning of the project to identify adhesives used for facing and their method of application. In addition conservators in the fields of easel paintings, paper and wall paintings were interviewed to find out which adhesives they used. Some initial testing was carried out on fragments of original paintings, at which stage pressure-sensitive adhesives, Paraloid B-72 and sodium alginate were ruled out for application on size-tempera. Initial testing was also used to work out which concentrations of the various adhesives were most suitable for further testing.

The following adhesives were selected for testing:

- *Protein glues*: gelatine 2.5%, SeaGel 100%, isinglass 2%, rabbit-skin glue 2%
- *Polysaccharide glues*: funori 1.5%, funori 2%, JunFunori 1.5%
- *Protein and polysaccharide mixtures*: funori 2%/gelatine 2% 1:1, funori 2%/isinglass 2% 1:1
- *Cellulose ethers*: Methocel A4C 1%, Methocel A4C 1.5%, Blanose 7M8SF 1%, Blanose 7M8SF 1.5%, Klucel E 5%, Klucel G 1.5%, Tylose MH300 1%
- *Acrylic dispersions*: Lascaux Medium for Consolidation 100%, Mowilith DMC2 DMC2 20%
- *Other synthetic adhesives*: Aquazol 200 8%, Aquazol 500 5%, Aquazol 50 in industrial methylated spirit (IMS) 10%, Aquazol 200 in IMS 10%, Aquazol 500 in IMS 10%, Beva 371 in Stoddard solvent
- *Volatile adhesives*: cyclododecane in IMS / petrolether 30-40 cyclododecane melt, menthol melt

Kozo paper and Eltoline tissue were identified as suitable facing papers during the preliminary testing.

Criteria and assessment of adhesives

The practical experimental research aimed to explore the potential of the identified adhesive materials for use as facing; it also considered the performance of adhesives in terms of both aesthetic and technical criteria. The adhesives were tested to assess the following criteria:

- The optical appearance of the adhesives, including changes during ageing.
- The interaction of the adhesives with different pigments.
- The mechanical properties of the adhesives, including changes during ageing.

Changes in the optical appearance of the adhesive samples were measured by colorimetry and glossimetry. Tensile testing was used to assess changes in mechanical properties. Additional properties such as pH and susceptibility to

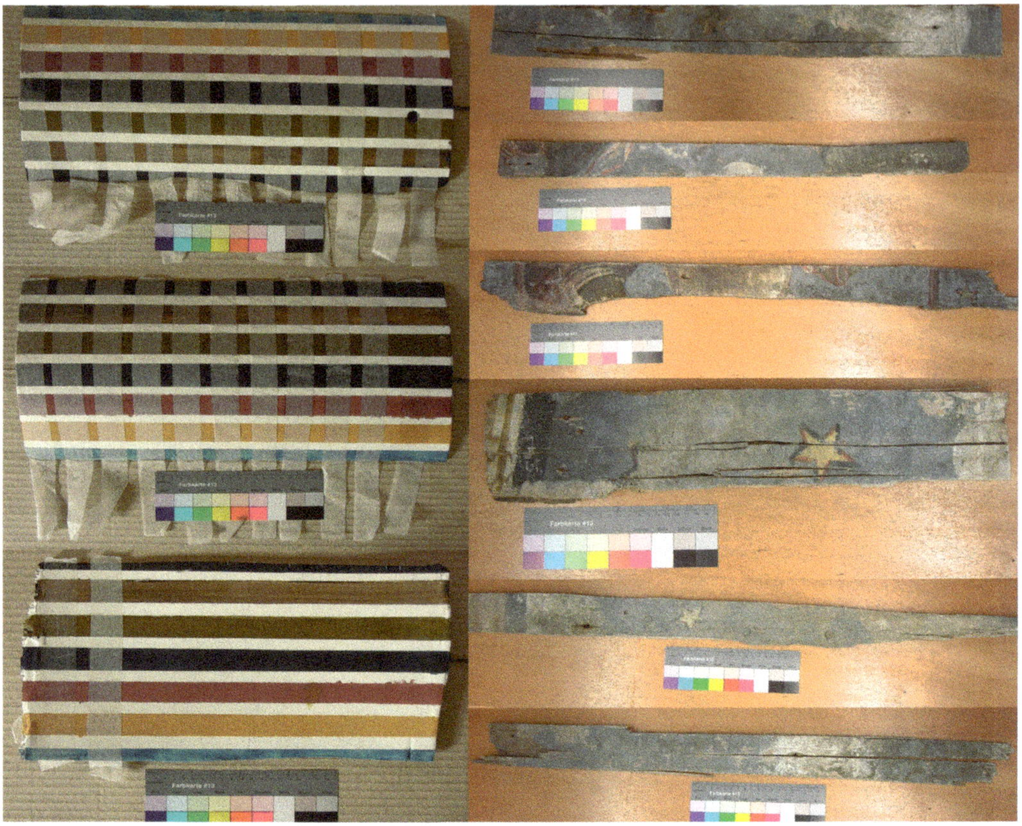

Figure 2 Replicas and original laths before testing. (Source: Chantal-Helen Thuer, Crown Copyright.)

microbiological attack were also considered but are not discussed in this paper.

Preparation of samples

Four different sets of samples were prepared with the adhesives for testing. The first set was prepared for tensile testing. Watercolour paper (200 g/m^2) was chosen as the substrate for the adhesive samples because of its reasonable strength, ability to spread applied liquids evenly and absorbency. The dimensions of each sample were 7 cm × 1.5 mm and those of the adhesive joint 3 × 1.5 mm. The adhesives were brushed onto the paper and the joined paper strips were weighted and dried for a week before testing.

A second set of samples was prepared for testing colour and gloss changes in the adhesives. Whatman filter paper 1 (circles 90 mm dia) was cut into strips of 1.5 cm width and dipped in a jar of a test adhesive up to a pencil mark then left to dry vertically. The advantages of using the filter paper were that it was of lab-grade quality, pH-neutral, absorbent, white and very matt.

The third and fourth sets of samples were prepared so that they replicated the practical application of a facing as closely as possible; in hindsight this proved useful, since it became evident after evaluation that the results on different materials produced varying results. For example adhesives that showed a questionable performance as a pure film on paper, or on size-tempera replicas, were found to work well on the original paint surfaces.

The third set of samples was applied to structural replicas which were approximately six months old. The replicas were made from old pine veneers and weathered plywood, painted with a ground of whiting bound in 5% rabbit-skin glue (Figure 2). Coloured stripes were applied to the panels using the pigments yellow ochre, iron oxide red, indigo, raw umber, iron oxide black, azurite and a mixed green (made of yellow ochre and indigo) bound in rabbit-skin glue. The replica was then faced using each of the adhesives, including the volatile ones, and Eltoline tissue in 3 cm strips across the coloured stripes. The faced replicas were also used for colorimetry to compare the adhesives' influence on the different pigments.

- *Panel No. 1, Eltoline*: gelatine, isinglass, rabbit-skin glue, funori 1.5%, funori 2%, JunFunori, funori/gelatine, funori/isinglass, Methocel A4C 1%, Methocel A4C 1.5%, Blanose 1%
- *Panel No. 2, Eltoline*: Blanose 1.5%, Klucel E, Klucel G, Tylose, SeaGel, Aquazol 200, 500, 50 IMS, 200 IMS, 500 IMS, Lascaux Medium for Consolidation (LMC)
- *Panel No. 3, Eltoline*: Mowilith, Beva 371, cyclododecane in solvent, menthol melt, cyclododecane melt

A final set of samples was applied to fragments of original ceiling panels. These were faced with two sets of strips, one with Eltoline tissue and the other with kozo tissue, and each of the adhesives. Original ceiling panel fragments with indigo paint layers and gilt, red-outlined stars were chosen because the comparably uniform paint made comparison between the different adhesives easier than it would have been on a highly coloured surface.

- *Panel No. 1, Eltoline*: gelatine, isinglass, rabbit-skin glue, funori 1.5%, funori 2%, JunFunori, funori/gelatine, funori/isinglass
- *Panel No. 2, Eltoline*: Methocel A4C 1%, Methocel A4C 1.5%, Blanose 1%, Blanose 1.5%, Klucel E, Klucel G, Tylose, SeaGel
- *Panel No. 3, Eltoline*: Aquazol 200, 500, 50 IMS, 200 IMS, 500 IMS, Lascaux Medium for Consolidation, Mowilith, Beva 371
- *Panel No. 4, Eltoline*: cyclododecane in solvent, cyclododecane melt, menthol melt
- *Panel No. 5, kozo*: Aquazol 200, 500, 50 IMS, 200 IMS, 500 IMS, Lascaux Medium for Consolidation, Mowilith
- *Panel No. 6, kozo*: Methocel A4C 1%, Methocel A4C 1.5%, Blanose 1%, Blanose 1.5%, Klucel E, Klucel G, Tylose, SeaGel
- *Panel No. 7, kozo*: gelatine, isinglass, rabbit-skin glue, funori 1.5%, funori 2%, JunFunori, funori/gelatine, funori/isinglass, Beva 371

During application of the facing samples to the replica paintings and original fragments, assessments were made of the adhesives' ease of use. Almost all adhesives in their chosen concentration were found to be easy to apply and were of appropriate viscosity to penetrate the tissues. JunFunori 1.5% and Methocel A4C 1.5% were an exception, being very viscous at that concentration and necessitating increased brushing for application. The dissolved cyclododecane was problematic during film formation, because the powdery, crystalline film redissolved immediately on repeated application making a thick, uniform formation nearly impossible to produce.

All the prepared samples, including the faced replicas and original fragments, were then stored for two weeks in cool, dark conditions to allow them to dry completely in the same climate.

After accelerated ageing the facings were removed from the replica paintings and original panels by brushing on the respective solvent (deionised water (50°C), white spirit or acetone), allowing it to soak the facing paper and then peeling it back with the application of more solvent if needed.

	Both Surfaces		Original Paint Surface								Replica Paint Surface					On Paper			
	Application		Facings		Surface changes		pH				Facings			Surface changes		Tensile Testing		Surface Change	
	General	Viscosity	adhesion on aged surface	Removal	Gloss	Colour	liquid adhesive	dried adhesive	aged adhesive	paint surface aged	Strong Pigment Influence on Facing Removal	adhesion on aged surface	Removal	Gloss	Colour	unaged	aged	Gloss	Colour
Gelatine 2.5% reference	(++)	(++)	(++/+)	(+/-)	(++)	(++)	5,7	5,61	5,65	6,55	G, A, I, U	(++)	(o)	(++)	(+/-)	(++)	(++)	(++)	(++)
Isinglass 2%	(++)	(++)	(+)	(+)	(++)	(++)	6,9	5,74	6,78	6,43	G, A, I, U	(++)	(o)	(++)	(++/+)	(++)	(++)	(++)	(++)
Rabbitskin Glue 2%	(++)	(++)	(+)	(+)	(++)	(++)	6,07	5,60	5,29	6,23	G, A, I, U	(++)	(-)	(++)	(+)	(+)	(++)	(++)	(++)
Funori 2%	(++)	(++)	(++)	(++)	(++)	(++)	6,73	6,92	8,00	6,58	(G)	(++)	(++)	(++)	(++)	(+)	(++)	(++)	(++)
Funori 1.5%	(++)	(++)	(++)	(++)	(++)	(++)	6,9	6,90	8,35	6,45	(G, U)	(++)	(++)	(++)	(++)	(+)	(++)	(++)	(++)
JunFunori 1.5%	(-)	(-)	(++)	(++)	(++)	(++)	7,11	7,12	7,97	6,58	(A, U)	(++)	(++)	(++)	(++/+)	(+)	(++)	(++)	(++/+)
Funori 2% / Gelatine 2% 1:1	(++)	(++)	(++)	(+)	(++)	(++)	6,85	6,19	6,44	6,50	A, I, U	(++)	(++)	(++)	(++/+)	(+)	(++)	(++)	(++)
Funori 2% / Isinglass 2% 1:1	(++)	(++)	(++)	(-/o)	(++)	(++)	7	6,12	7,18	6,41	R, G, I, U	(++)	(oo)	(++)	(++/+)	(++)	(+)	(++)	(++)
SeaGel 100%	(++)	(++)	(oo)	(++)	(+)	(-)	7,14	6,38	6,42	6,59		(oo)	(++)	(+)	(+/-)	(+)	(o)	(++)	(-)
Methocel A4C 1.5%	(-)	(-)	(-)	(++)	(++)	(++)	6,94	6,28	7,27	6,43	(U)	(++)	(++)	(++)	(+)	(++)	(+)	(++)	(++)
Methocel A4C 1%	(+)	(+)	(+)	(++)	(++)	(++)	6,03	6,90	7,77	6,52	(U)	(++)	(++)	(++)	(+)	(+)	(+)	(++)	(++)
Klucel E 5%	(++)	(++)	(oo)	(++)	(+)	(+/-)	6,75	6,70	7,43	6,56	(A, G, I)	(++)	(++)	(+)	(-)	(o)	(o)	(++)	(++)
Klucel G 1.5%	(+)	(+)	(oo)	(++)	(++)	(++)	7,24	7,19	8,16	6,57	(A, G, I)	(++)	(++)	(++)	(++/+)	(-)	(o)	(++)	(++)
Blanose 7M8SF 1.5%	(+)	(+)	(-)	(++)	(+)	(++)	7,32	7,18	7,61	6,42		(++)	(++)	(+)	(+)	(++)	(++)	(++)	(++)
Blanose 7M8SF 1%	(+)	(+)	(+)	(+/-)	(+)	(++)	7,4	7,20	7,49	6,65		(++)	(++)	(+)	(++/+)	(++)	(+)	(++)	(++)
Tylose MH300 1%	(+)	(+)	(+)	(++)	(++)	(++)	6,94	7,49	7,64	6,55	G, A, U	(++)	(++)	(++)	(++/+)	(++)	(+)	(++)	(++)
Aquazol 200 8%	(++)	(++)	(o)	(++)	(+)	(+)	6,22	7,13	7,37	6,53		(++)	(++)	(+/-)	(+)	(++)	(+)	(++)	(++)
Aquazol 500 5%	(+)	(+)	(-/o)	(++)	(+)	(+)	6,33	7,37	7,82	6,36		(++)	(++)	(+/-)	(++)	(+)	(+)	(++)	(++)
Aquazol 50 10% IMS	(++)	(++)	(o)	(++)	(+)	(+/-)	5,59	7,31	7,35	6,41		(+)	(++)	(-)	(+/-)			(++)	(++)
Aquazol 200 10% IMS	(++)	(++)	(-/o)	(++)	(++)	(+/-)	5,65	7,09	7,17	6,28		(+)	(++)	(-)	(-)			(++)	(++)
Aquazol 500 7% IMS	(+)	(+)	(-/o)	(++)	(++)	(+/-)	5,81	7,19	7,14	6,41		(+)	(++)	(-)	(+/-)			(++)	(++)
BEVA 371 Lascaux thin foil																(+)	(+)		
BEVA 371 30%	(+) solvent!	(++)	(+)	(+)	(++)	(+/-)	4,94	5,83	5,80	6,72	(O)	(++)	(+)	(++)	(+/-)			(++)	(-/o)
LMC 100%	(++)	(++)	(++)	(oo)	(-)	(-/o)	8,02	5,43	5,68	6,68	A, U, B, (rest)	(++)	(oo)	(-)	(oo)	(-)	(++)	(++/+)	(o)
Mowilith DMC2 20%	(++)	(++)	(++)	(-)	(-)	(+/-)	4,2	5,93	5,45	6,44		(++)	(o)	(-)	(-/-)	(-)	(+)	(++)	(+/-)
Cyclododecane dissolved	(-) brush!	(++)	(o) sublimated	n/a	(+)	(++)		7,28	n/a	6,61		(o) sublimated	n/a	(+)	(++)			(++)	(++)
Cyclododecane melt	(+) brush!	(+)	(++) sublimated	n/a	(++)	(++)		7,25	n/a	6,71		(++) sublimated	n/a	(++)	(++)			(++)	(++)
Menthol melt	(+) brush!	(+)	(++) sublimated	n/a	(++)	(++)		7,27	n/a	6,44		(++) sublimated	n/a	(++)	(++)			(++)	(++)

Performance Grading: (++) very good, (+) good, (-) medium, (o) poor, (oo) unsuitable

Pigment Code: W chalk white, O goldochre, R ironoxide red, G green, A azurite, I indigo, U raw umber, B ironoxide black

Figure 3 Comparison chart of all tested adhesives for all test sequences. (Source: Chantal-Helen Thuer, Crown Copyright.)

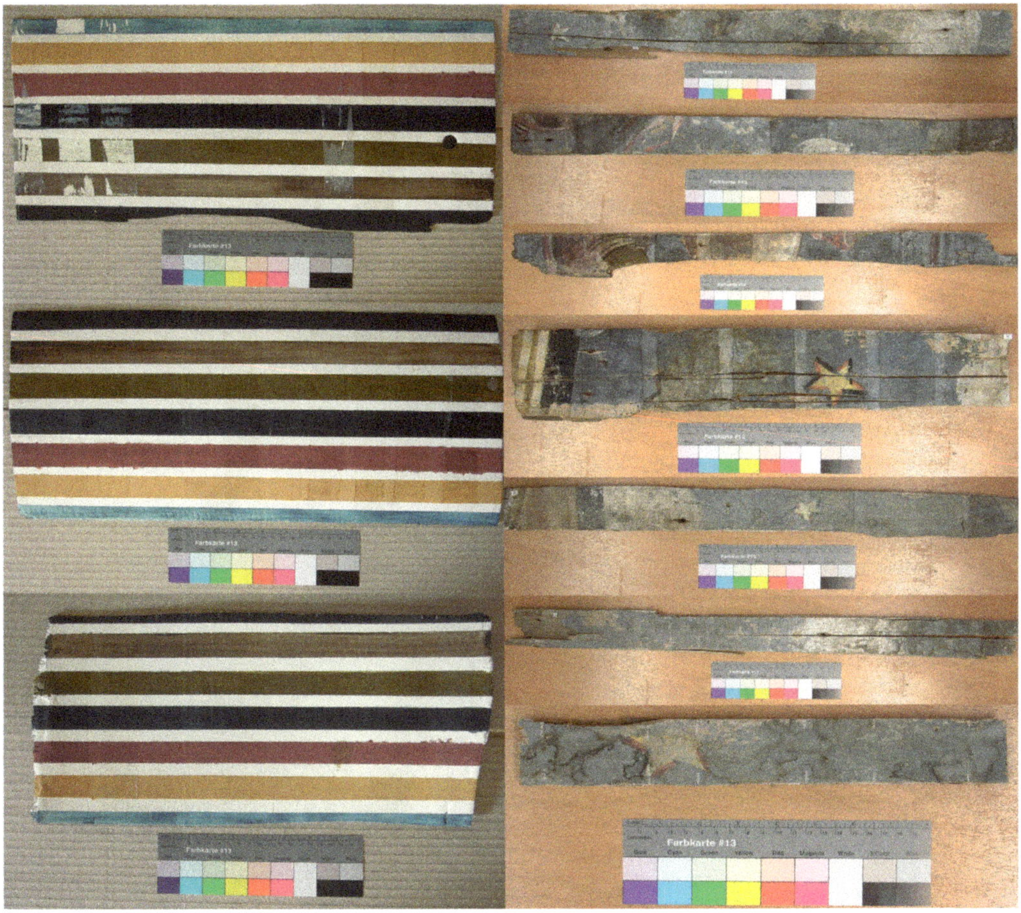

Figure 4 Replicas and original laths after removal of the facing tissue. (Source: Chantal-Helen Thuer, Crown Copyright.)

Accelerated ageing

While it was acknowledged that accelerated ageing cannot be directly equivalent to decades of natural ageing, it was considered that the results produced by accelerated ageing could still provide indicators of what may happen over time, showing possible dangers and weaknesses.

An overview of the environmental conditions to which painted ceilings are exposed was collected using data loggers at Historic Scotland and the National Trust for Scotland properties. The data indicated that the tested adhesives would have to withstand relatively high, long-term relative humidity (RH) in the range of 60–95% RH and temperatures fluctuating between 5 and 25°C. Accelerated ageing of the samples was therefore adapted in an attempt to replicate Scottish environmental conditions.

Due to technical reasons the ultraviolet (UV) light ageing had to be separated out from the weathering sequence of testing. The UV ageing was executed with two lamps with two BLB tubes (peak at 365 nm) and Vilber-Lourmat UV-A tubes (T-20.M), peaking at 312 nm. The overall exposure time was 11 days. To monitor the climate in the room a Tinytag Ultra 2 logger was installed.

The accelerated weathering was undertaken in a Gallenkamp Fitotron Plant Growth

Chamber, Model SGC097.PPX.101. The accelerated ageing cycle was programmed to use the weather extremes of each Scottish season as a segment within the cycle. A worst case scenario cycle, with amendments due to the chamber's performance, was spring 8°C/70% RH, summer 23°C/50% RH, autumn 12°C/80% RH and winter 3°C/95% RH, running three hours each with lights on and off every half an hour to emulate night and day, bringing the full cycle to 12 hours.

Testing and discussion of results

To gain an overview of the adhesives' performance the data from the testing are assembled in chart form in Figure 3.

Facing removal and adhesive–pigment interactions

It has been shown that pigments can influence the behaviour of binding media. For example, research has revealed different behaviours and denaturation of protein binders in the presence of certain pigments (Fiedler 2001: 99). This could suggest that an adhesive for matt paint that is introduced into the paint layer as a 'replacement binding medium' might react differently with different pigments. The adhesive could change chemical structures in the paint layer during ageing or pigments could affect the stability of an adhesive.

Comparison of the test areas after ageing on the original paint surfaces and on the structural replicas revealed differences, suggesting that using replicas instead of original material does not necessarily produce results applicable to the original (Figure 4).

Facings on the replicas adhered very well after ageing, apart from the SeaGel facing, which had fallen off due to the total failure of the adhesive. Conversely, the facings on the original panels showed signs of failure in the form of ripples and pronounced delamination, SeaGel, Klucel E, Klucel G, Aquazol 200 and Aquazol 50 IMS being complete failures. Significant delamination of the facing was observed with Methocel A4C 1%, Blanose 7M8SF 1.5%, Aquazol 500 5%, Aquazol 200 10% IMS and Aquazol 500 7% IMS. Only funori and JunFunori were still able to hold the tissue flush to the paint surface of the original panels.

When removing the facings the same tendencies could be observed on both original and replica, however the results were much more pronounced on the replica. The gelatine and funori/isinglass facings became more difficult to remove, taking a visible amount of paint with them. On the replicas these protein glues damaged the azurite, indigo, green and raw umber sample stripes to the point where the indigo and green stripes stuck completely to the facing, and azurite and raw umber were partially adhered. Most other adhesives, when applied to azurite, indigo, green and raw umber exhibited reduced solubility. This may support findings by Fiedler (2001: 56) that some pigments can cause a significant decrease in solubility. According to that study and that of Schilling and Khanjian (1996: 127, 128, 138), proteinaceous materials are influenced by certain pigments as their reactive hydrophilic amino acids react with the pigment, decreasing solubility as these groups are no longer available. The differences in the amounts of those amino acids in each protein glue (Schilling and Khanjian 1996: 141) and the availability of hydrophilic groups, could account for variations in solubility. However on the original paint layer indigo seemed to pose less of a problem than on the replicas.

After performing colorimetry it became evident that the degree of discoloration was also influenced by the pigment used and the surface to which an adhesive was applied (Figure 5). Considering this, it seems advisable that consolidants should be tested on different pigments before application. However it should be noted that the level of damage to the replicas was not observed on the original panels.

ΔE^*_{ab}	**Evaluation Chalk White**	**Adhesives**
0-1	No colour difference	Menthol, CDD both, Aquazol 200 IMS, Aquazol 200, Tylose, Blanose both, Klucel both, Funori 1.5%
1-2	Minimal colour difference ("skilled eye")	Mowilith, LMC, Aquazol 500 IMS, Aquazol 50 IMS, Aquazol 500*, Methocel both, Funori/Isinglass, Funori/Gelatine, Funori 2%, Gelatine
2-4	Visible colour difference	Isinglass, Rabbitskin, JunFunori*
4-5	Great colour difference	BEVA, SeaGel*
> 5	Extreme colour difference	-
Void		-

ΔE^*_{ab}	**Evaluation Gold Ochre**	**Adhesives**
0-1	No colour difference	CDD solv., Aquazol 500, Klucel G, Funori/Gelatine, Funori 1.5%
1-2	Minimal colour difference ("skilled eye")	CDD melt, BEVA, Aquazol 200, Tylose, Blanose both, Klucel E, Methocel both, Funori/Isinglass, Funori 2%, Rabbitskin, Isinglass, Gelatine
2-4	Visible colour difference	Menthol, SeaGel, JunFunori
4-5	Great colour difference	Mowilith, LMC, Aquazol 500 IMS*, Aquazol 50 IMS
> 5	Extreme colour difference	Aquazol 200 IMS
Void		--

ΔE^*_{ab}	**Evaluation Ironoxide Red**	**Adhesives**
0-1	No colour difference	Cyclododecane both, Aquazol 500, Blanose 1%, Funori/Isinglass, JunFunori, Funori 2%, Isinglass
1-2	Minimal colour difference ("skilled eye")	Menthol*, Mowilith, LMC, Tylose, Blanose 1.5%, Funori/Gelatine*, Funori 1.5%, Rabbitskin
2-4	Visible colour difference	BEVA, Aquazol IMS all, Aquazol 200, Klucel both (G*), Methocel both, SeaGel, Gelatine*
4-5	Great colour difference	-
> 5	Extreme colour difference	-
Void		-

ΔE^*_{ab}	**Evaluation Green**	**Adhesives**
0-1	No colour difference	Menthol, Cyclododecane melt, Aquazol 500 IMS, Aquazol 500, Tylose, Methocel 1.5%, Funori 1.5%
1-2	Minimal colour difference ("skilled eye")	Cyclododecane solv.*, Mowilith, LMC*, Aquazol 200 IMS, Aquazol 50 IMS, Aquazol 200, Blanose 1%, Klucel G, SeaGel, Funori/Gelatine, JunFunori, Funori 2%
2-4	Visible colour difference	Blanose 1.5%, Klucel E, Methocel 1%*
4-5	Great colour difference	-
> 5	Extreme colour difference	BEVA
Void		Funori/Isinglass, Rabbitskin, Isinglass, Gelatine

ΔE^*_{ab}	Evaluation Azurite	Adhesives
0-1	No colour difference	Cyclododecane melt, JunFunori
1-2	Minimal colour difference ("skilled eye")	Menthol, Cyclododecane solv., Aquazol 200, Blanose both, Klucel G*
2-4	Visible colour difference	Mowilith, BEVA, Aquazol 50 IMS, Tylose, Klucel E, Methocel both (1%*), SeaGel, Funori/Gelatine, Funori 2%
4-5	Great colour difference	Aquazol 200 IMS, Rabbitskin, Gelatine
> 5	Extreme colour difference	LMC, Aquazol 500 IMS, Aquazol 500*, Funori 1.5%
Void		Funori/Isinglass, Isinglass
ΔE^*_{ab}	**Evaluation Indigo**	**Adhesives**
0-1	No colour difference	Klucel G, Methocel both, Funori/Gelatine
1-2	Minimal colour difference ("skilled eye")	Menthol, Cyclododecane both (melt*), BEVA, Aquazol 500 IMS, Aquazol 500, Tylose, Blanose both (1%*), SeaGel, JunFunori, Funori both
2-4	Visible colour difference	Mowilith, Aquazol 50 IMS, Aquazol 200, Klucel E, Funori/Isinglass*, Rabbitskin
4-5	Great colour difference	Aquazol 200 IMS
> 5	Extreme colour difference	LMC*
Void		Isinglass, Gelatine
ΔE^*_{ab}	**Evaluation Raw Umber**	**Adhesives**
0-1	No colour difference	Menthol, Aquazol 500, Klucel G, Funori both, Rabbitskin
1-2	Minimal colour difference ("skilled eye")	Cyclododecane both, Mowilith*, BEVA, Aquazol 500 IMS*, Aquazol 50 IMS, Aquazol 200, Methocel 1%
2-4	Visible colour difference	LMC, Aquazol 200 IMS, Tylose*, Blanose 1.5%, Klucel E, Methocel 1.5%, SeaGel, JunFunori, Isinglass, Gelatine
4-5	Great colour difference	Funori/Gelatine
> 5	Extreme colour difference	Blanose 1%, Funori/Isinglass
Void		-
ΔE^*_{ab}	**Evaluation Ironoxide Black**	**Adhesives**
0-1	No colour difference	all others
1-2	Minimal colour difference ("skilled eye")	Funori both
2-4	Visible colour difference	LMC
4-5	Great colour difference	-
> 5	Extreme colour difference	-
Void		-

* bordering onto the lower category

Figure 5 Delta_E*ab evaluations for the differently coloured areas on the replicas after the removal of the facing tissue. (Source: Chantal-Helen Thuer, Crown Copyright.)

Optical changes in adhesives

Colorimetry

Colorimetry was used to assess two different types of colour change: that caused by the application of the adhesive, and the colour change that appeared during ageing. The colorimetry was recorded in the CIE-L*a*b* colour space with a Konica Minolta Chroma Meter CR-40.

To measure the colour differences between the adhesives and paint layers, the Δ values, the 'absolute' values, were recorded (BS 3483-A7:2006; BS 6923:1988). The Δ values describe the deviation of the three colour values: lightness (ΔL*), red/green (Δa*) and blue/yellow (Δb*), from a reference to a newly measured sample. The complete colour deviation, ΔE^*_{ab}, is the total amount of colour change but does not indicate the direction of colour change. The results can be interpreted using the values in Table 1 as a guide (see also Figure 6).

Table 1 Results of the measurement of the colour differences.

ΔE^*_{ab}	Evaluation
0–1	No colour difference
1–2	Minimal colour difference ('skilled eye')
2–4	Visible colour difference
4–5	Great colour difference
>5	Extreme colour difference

Results from filter paper substrate samples

Most of the adhesives showed minimal to no colour difference with ΔE^*_{ab} values between 0 and 2. The exceptions were the Lascaux Medium for Consolidation and Beva 371 with values of ΔE^*_{ab} in the range of visible colour difference. The next adhesive on the scale of colour change was the SeaGel, which showed significant change after UV ageing but reverted to a ΔE^*_{ab} in the minimal colour difference range after accelerated ageing.

Yellowing (positive Δb*) was expected for many adhesives after ageing, but most of the samples exhibited a decrease in yellowness. However as with ΔE^*_{ab} values, most of the changes for Δb* were minimal, the majority being below 1, so would not be perceptible. Values for Beva 371 and Lascaux Medium for Consolidation were recorded, which indicated visible yellowing. The results for Beva 371 support research at the Canadian Conservation Institute, which had shown evidence of the yellowing of Beva 371 (Down *et al.* 1996). SeaGel showed significant yellowing on application, which increased after UV ageing, but this reverted to a minimal level, below 2, after climate ageing.

The colour shift on the red-green plane (Δa*) was negligible for most adhesives with all results indicating minimal colour difference.

Results from samples on replicas and original fragments

Darkening of the paint layer can be a sign of adhesive saturation in a porous paint. During testing on the replicas, darkening also seemed to depend on a pigment's absorption ability. The most notable darkening of the tested adhesives was observed with Lascaux Medium for Consolidation and Mowilith DMC2. The colorimetry showed that darkening was detected with most adhesives, dependent on the pigment type, accounting for a great part of ΔE*ab. Generally the synthetic adhesives were most problematic on the original paint layer causing surface saturation. This was observed during the testing but discoloration could not be verified with the colorimeter due to the highly uneven appearance of the paint surface.

The second biggest influence was the amount of yellowing. Comparing all ageing performances by the overall value for colour difference, ΔE*ab, for each pigment stripe Lascaux Medium for Consolidation is unsuitable, while gelatine, SeaGel, Klucel E, Aquazol in IMS, Beva 371 and Mowilith DMC2 can cause perceptible colour changes. It should be noted that on the paper samples, only SeaGel, Beva 371, Lascaux Medium and Mowilith DMC2 showed a distinct change.

Discoloration of other adhesives had been pronounced after the UV ageing but often reverted to a lower discoloration after climate ageing.

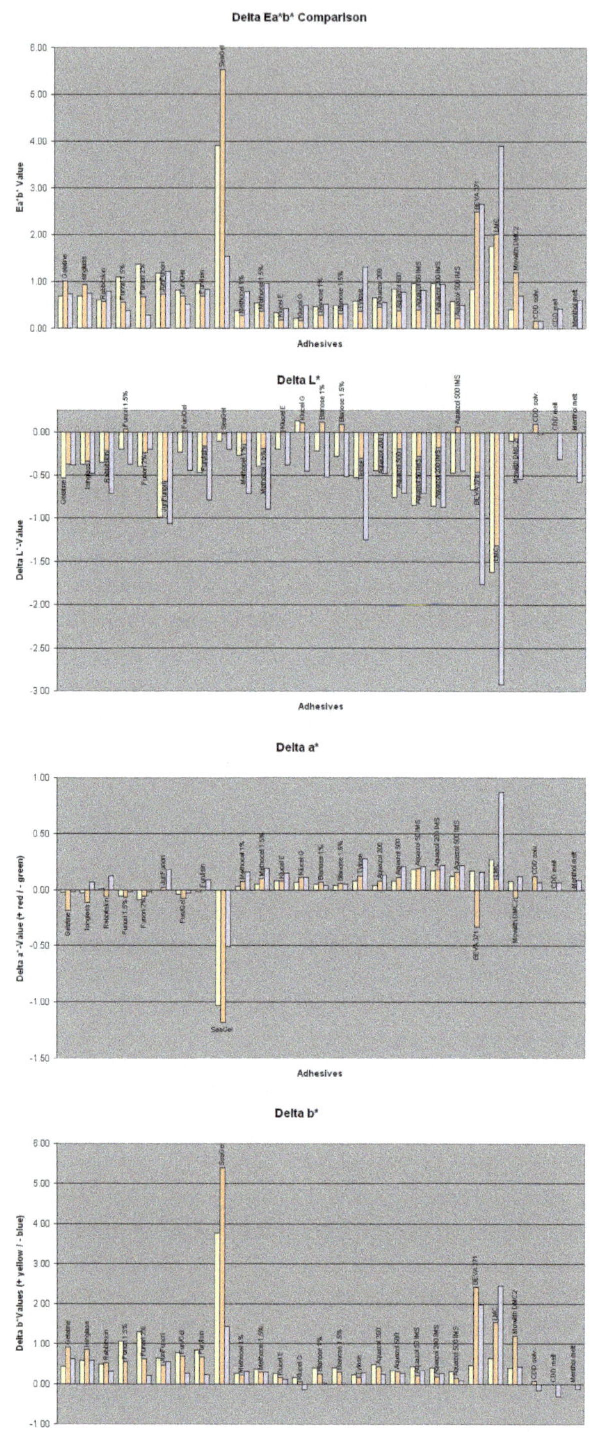

Figure 6 Colorimetry results showing performance before (yellow), after UV ageing (red) and after climate ageing (blue). (Source: Chantal-Helen Thuer, Crown Copyright.)

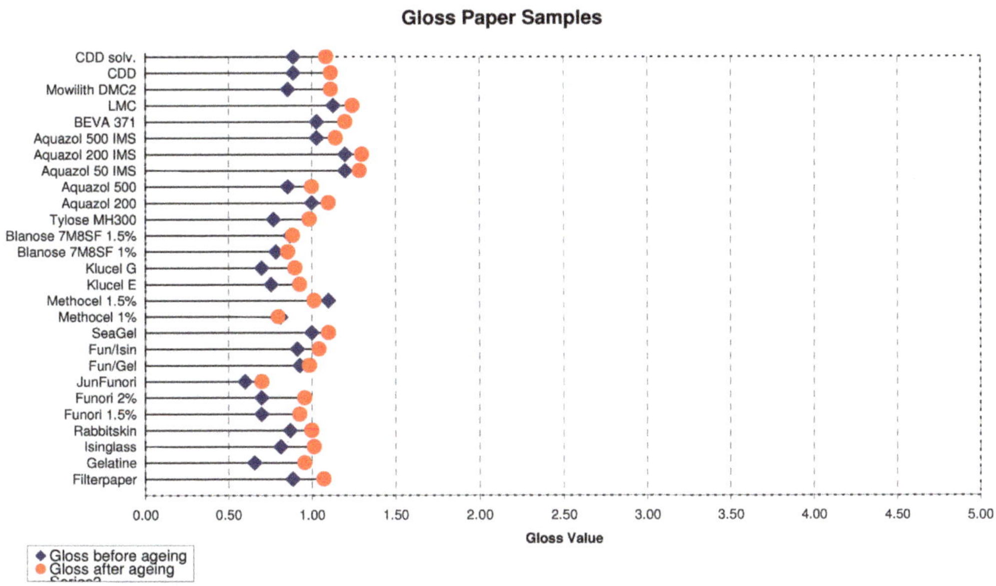

Figure 7 Glossimetry results. 1 is generally a very low value meaning not much gloss. (Source: Chantal-Helen Thuer, Crown Copyright.)

Glossimetry

Glossimetry was applied with an Elcometer 402 (by Elcometer Instruments Ltd) to the structural replicas and the filter-paper samples dipped in the adhesive. The gloss was measured at an angle of 85° for more accuracy on the matt surface.

Results on the test samples showed that an increase in gloss from application of the facing adhesives was a minor problem. It was found that there was hardly any difference in the matt appearance of the filter paper where the adhesive had been applied. The old paint layers on the fragments of original ceiling panels became glossier with the application of Lascaux Medium for Consolidation and Mowilith DMC2 (Figure 7). The replica paint surface was more susceptible to gloss and Aquazol, especially in IMS, also produced increased gloss. It was notable that certain pigments were less glossy than others, as more porous pigments such as chalk and ochre absorbed more facing adhesive from the surface. The more pronounced gloss on the replicas was possibly due to the less denaturated binder, which may have prevented absorption of the facing adhesive compared to the filter paper samples and original painting fragments.

Mechanical changes of adhesives

Tensile testing

Tensile testing was carried out on the samples of adhesive applied to a watercolour paper substrate. Comparison was made between the tensile testing results of unaged and aged adhesive samples. The aged samples were kept at room temperature for a week in the studio before they were tested. Fmax (maximal force in newtons) strain at max (in mm) and the elastic modus Emod were recoded and compared for each sample. A Zwick/Roell ProLine Z050 TN material testing machine with 5 kN clamps was used at a speed of 1mm/s.[1]

Results

Comparing the tensile testing data for strain, tensile force and elastic modus, before and

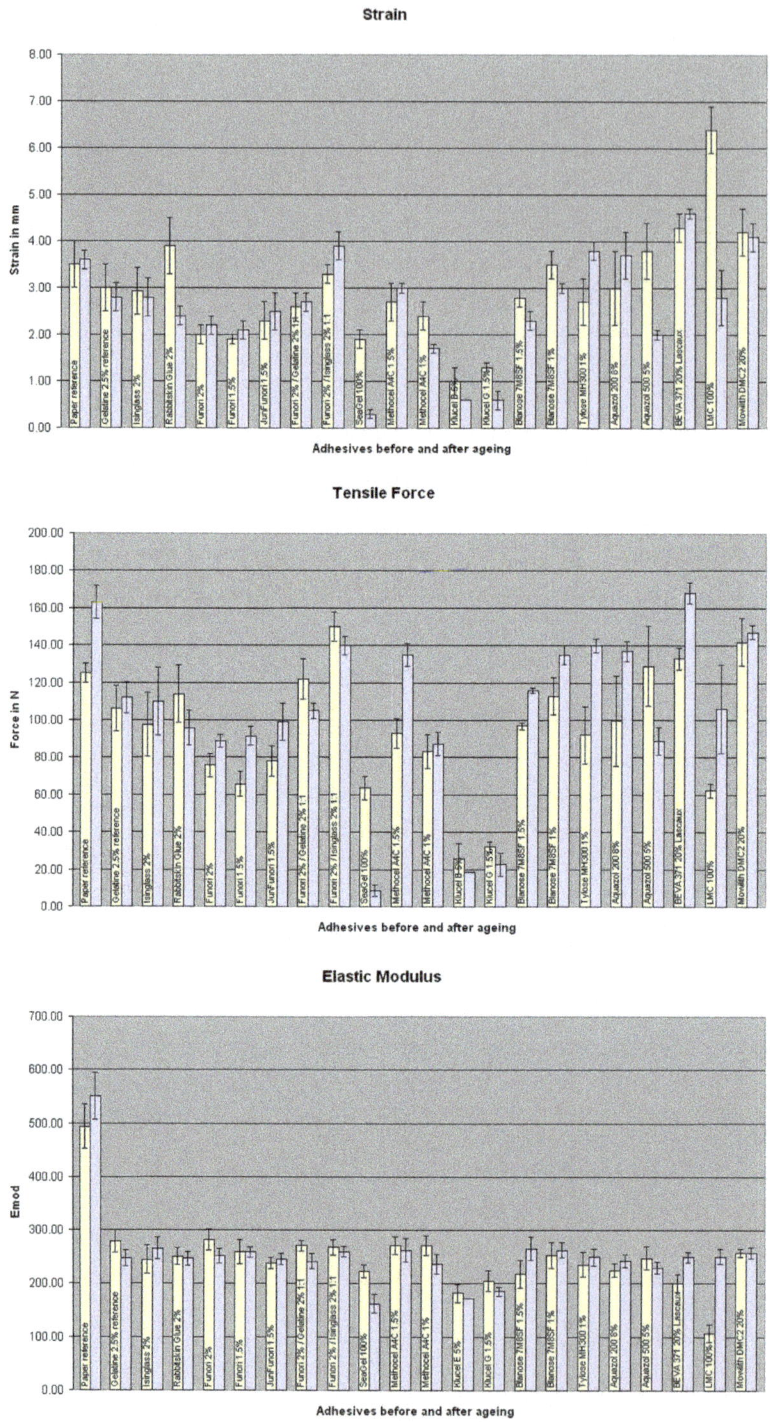

Figure 8 Tensile testing charts showing performance before and after ageing. (Source: Chantal-Helen Thuer, Crown Copyright.)

after accelerated ageing, it seems that the flexibility–brittleness ratio was mostly unchanged. This was surprising as the non-crystalline adhesives were generally expected to show greater differences in the elastic modulus (Figure 8). Since SeaGel is basically a protein glue, which is considered brittle, it is thought that the increased flexibility could be related to the additives that take up 3.5% to 13% solid contents in the liquid. The increase in brittleness for Blanose 1.5%, Beva 371 and Lascaux Medium for Consolidation could be ascribed to some cross-linking in the polymer structure. Since the Lascaux Medium for Consolidation was thoroughly tested by Lascaux during its development, and no chemical changes were reported during accelerated ageing, this result was unexpected for this adhesive (Hedlund and Johansson 2005: 437) .

Changes due to ageing were more obvious in this test series when considering the changes in plasticity and tensile strength. A loss in plasticity was observed for most adhesives. This can be attributed to the ageing processes of cross-linking and chain scission, activated by UV light, oxidisation and hydrolysis. These reactions restrict the movement of the polymer chains within the adhesive and reduce the amount of strain it can take. The variety of results between different concentrations of the same adhesive was surprising.

The amount of tensile strength recorded for the different adhesives reflects the ageing processes even more. Those adhesives with an increased strength after ageing would probably have experienced a greater amount of cross-linking, making the larger number of bonds difficult to break. The adhesives with decreased strength after ageing may have experienced more chain scission, making adhesion failures easier. This would accord with the findings of Geiger and Michel (2005: 198).

Accelerated weathering and UV ageing seems to have had the least influence on the tensile properties of the protein glues. SeaGel should be considered separately because of the multitude of additives the other tested protein glues do not have. The synthetic adhesives seem to have been affected the most by the accelerated ageing.

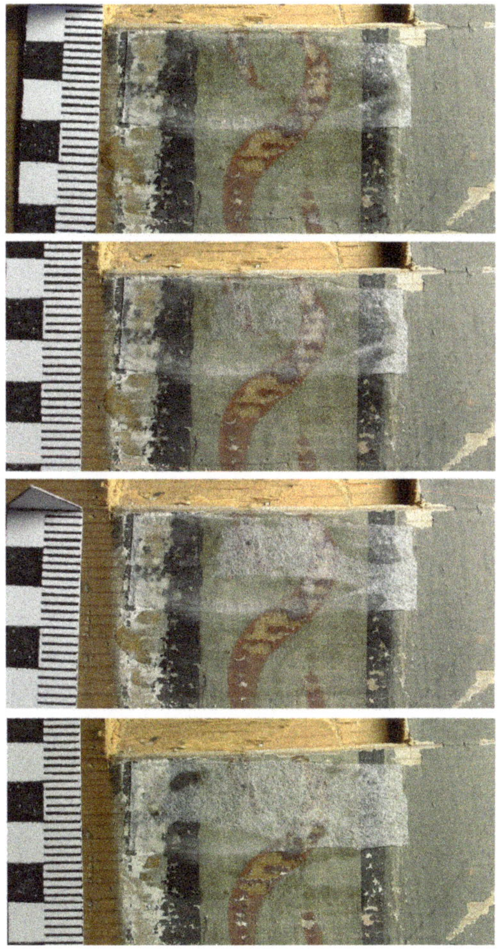

Figure 9 Sublimation of cyclododecane melt, applied with and without tissue: area new, after 1 day, 2 days and a week. (Source: Chantal-Helen Thuer, Crown Copyright.)

By comparing the tensile testing results to the facing delamination on the samples, it is evident that although ageing and cross-linking in the adhesives increased their tensile strength, especially in the case of Beva 371, the facings still had a tendency to fail before any damage could be done to the paint layer. This failure seems to have been caused by a reduction in plasticity because the flexibility of the adhesives hardly changed at all. The adhesives with low tensile strength and plasticity were also those where the facings delaminated the most, namely SeaGel, Klucel E and Klucel G. The significant failure of Aquazol in IMS may have been caused by an inherent change in tensile properties or by its greater penetration into the porous original paint layer.

Practical applications

Acrylates

After considering the performances of all the adhesives in the different testing sequences it was found that acrylic emulsions and solvent-based acrylates were generally not suitable as facing adhesives for size-tempera paintings because of poor removability, effects on gloss and discoloration. Although Lascaux Medium for Consolidation was designed for matt paint layers similar to the painted wooden interiors in Scotland, from the results of this research it would appear to be best suited as a consolidant when applied underneath flakes without leaving residues on the surface. As a facing adhesive and consolidant through facings it was found to be unsuitable. The results for Beva 371 showed that it would be suitable under certain circumstances and for specific pigments. It has been used successfully on a board-and-beam ceiling in a Historic Scotland estate, where it accorded with previous treatments. However the restricted use of aqueous adhesives after Beva 371 treatment should always be considered.

Aquazol

Aquazol's performance was dependent mainly on the chain lengths and solvents used. Its major problems were found to be poor facing adhesion and darkening. However, the facing adhesion and ageing characteristics of the water-bound adhesive were slightly better than those of the IMS-bound one. The results indicated that Aquazol 200 8% could prove suitable in some circumstances and for certain pigments. Since all Aquazol types are soluble in a wide range of solvents outside those tested, trials with another slower evaporating solvent might yield better results for gloss, darkening and facing adhesion.

Volatile adhesives

As facing adhesives, volatile adhesives performed well and the melts were found to be suitable despite possible problems with the application method. The addition of a facing tissue was found to help reduce the sublimation time and improve surface protection. This could be especially useful in situations where paint flakes could easily be dislodged when the supporting adhesive sublimates. The limited time before sublimation makes them most suitable for transport or short-term application (Figure 9).

Cellulose ethers

Of the cellulose ethers both Klucel types were found to be unsuitable because of darkening and poor facing adhesion. Of the remaining adhesives in this category Methocel A4C 1%, Blanose 7M8SF 1% and Tylose MH300 1% were preferable, showing the strongest facing adhesion and causing the least surface changes, although consideration should be given to pigment influences on reversibility. They could also prove suitable as consolidants, applied through a facing.

Natural adhesives

Among the natural adhesives, funori and JunFunori showed the best overall performance of all adhesives along with the volatile adhesives, although they are considered to be 'weak' adhesives. After ageing their facings were the only ones that were still flush with the surface on all test substrates and there was no gloss, discoloration or darkening. The only discernible change was that the original paint layer became slightly lighter – perhaps due to a cleaning effect. The JunFunori performed as well as natural funori, but was more difficult to apply due to its viscosity and low penetration. Further testing in the field would be required to ascertain whether both adhesives are strong enough to work as a consolidant as well.

The three protein glues performed less well than most of the other adhesives, gelatine being the most difficult to re-solubilise. Under certain circumstances isinglass could be equally as problematic in its removal as gelatine. Rabbit-skin glue had slightly more favourable ageing characteristics than isinglass. The funori/isinglass mixture was found to be unsuitable due to its poor removability. The funori/gelatine mixture performed better, although it might be more suitable as a consolidant applied through a facing. However gelatine has some benefits as a consolidant, as it generally keeps the paint layers intact and remains mostly unchanged over a long period of time.

Case-related recommendations

Long-term facings

From the tested adhesives, funori (1.5% / 2%) and JunFunori 1.5% would be best suited for long-term facings, especially on installed walls and ceilings; they perform exceptionally well and are readily removable after accelerated ageing. After removal no tide lines were visible and the only evidence of treatment was a very slight increase in brightness. Aqueous treatments would be possible after the application of both adhesives.

Temporary facings

As facing adhesives for short-term transport and storage, volatile adhesives in the form of melts could be useful despite the added care needed to apply them and consideration of their time-limited nature. Cyclododecane and menthol both left no visible evidence on any test surface once they had sublimated. The sublimation time was prolonged with the use of facing tissue. Menthol might pose difficulties in badly ventilated areas due to its strong odour and it was not possible to test whether any essential oils remained on the surface during this research.

For transport, the cyclododecane melt would provide a thick, waxy layer that could protect tented paint flakes from abrasion and damage. Due to its sublimation process the time it stays on the surface is determined by the thickness of its layer, which can vary from weeks to a few months. The use of a sublimating adhesive could prove problematic if the term of temporary facing was unexpectedly extended, for example if a conservation treatment was delayed. Sublimation could potentially be inhibited by covering the layer with a non-breathable material or adhesive cover layer, for example a cellulose ether (Brückle 1990), however this was not tested.

Facing adhesives with consolidating effect

If an adhesive is chosen as a consolidant to be applied through a facing tissue, as well as gelatine and isinglass, the cellulose ethers Methocel A4C 1%, Blanose 7M8SF 1% and Tylose MH300 1% could be a good choice. It should be noted that the adhesives' ageing performance has not been tested as a consolidant only, and the choice has been based on experience in the field and their performance as facing adhesives rather than as consolidants.

Adhesives found unacceptable for above applications

Based on this research the funori/isinglass mixture, SeaGel, Klucel E /G, the synthetic adhesives Aquazol, Lascaux Medium for Consolidation and Mowilith DMC2 were considered unsuitable for applying facings to size-tempera paint.[2]

Materials and suppliers

- Technical gelatine (Kremer)
- Isinglass (Kremer)
- Rabbit-skin glue (Cornelissen & Son)
- SeaGel (Conservator's Products Company / Conservation Resources Ltd, discontinued and replaced with Fish Gelatine, www.fish-gel.com/Newsite/Blank_2.html) – fish gelatine (~13%), propylene glycol (~3%), potassium sorbate (~0.5%) and water (~83.5%)

- Funori – polysaccharide mucilage from the seaweed *Gloiopeltis furcata*
- JunFunori – refined version of Funori by Lascaux
- Sodium alginate – sodium salt (NaC6H7O6) of alginic acid extracted from the brown algae *Laminaria hyperborea*
- Methocel A4C (Dow Chemicals) – methyl hydroxyl propyl cellulose
- Blanose 7M8SF (Aqualon / Hercules) – sodium salt of carboxy methyl cellulose
- Klucel G (Conservation Resources) – hydroxy propyl cellulose
- Klucel E (Conservation Resources) – hydroxy propyl cellulose, less viscous than G-type
- Tylose MH300 (Lascaux) – methyl hydroxyl ethyl cellulose
- Paraloid B72 (Lascaux) – methyl / ethyl methacrylate copolymer
- Lascaux Medium for Consolidation (Kremer / Lascaux) – aqueous dispersion of an acrylic copolymer, various additives (~3%) and solvents (~2%)
- Lascaux 360HV (Lascaux) – methyl methacrylate, butyl acrylate, thickened with acrylic butyl ester
- Mowilith DMC2 (Lascaux) – copolymer of vinyl acetate and maleic acid dibutyl ester
- Aquazol 50/200/500 (Polymer Chemistry Innovations Inc. (PCI)) – poly-2-ethyl-oxazoline (C5H9NO), different molecular weight
- Beva 371(Lascaux) – ethylene vinyl acetate
- Cyclododecane (Kremer / Hangleiter) – non-polar, saturated, alicyclic hydrocarbon
- Menthol (Kremer) – menthanol-3 / 1-methyl-4-isopropyl-cyclohexanol-(3)

Notes

1. The testing was done by Lynn Chalmers from the Materials and Manufacturing Research Group at Napier University, initiated by Alan Davidson.
2. The full research report is available at www.historic-scotland.gov.uk/technicalpaper11.pdf (accessed 5 September 2012).

References

Brückle, I., Thornton, J., Nichols, K. and Strickler, G. (1990) 'Cyclododecane: technical note on some uses in paper and objects conservation', *Journal of the American Institute for Conservation* 38(2): 162–75.

Down, J.L., MacDonald, M.A., Tetreault, J. and Williams, R.S. (1996) 'Adhesive testing at the CCI: an evaluation of selected poly(vinyl acetate) and acrylic adhesives', *Studies in Conservation* 41(1): 19–44.

Fiedler, I. (2001) *Anorganische Biochemie und Kunst – Zur Stabilität proteinhaltiger Matrices in Gegenwart anorganischer Pigmente*, PhD dissertation, Eberhard - Karls - University Tübingen.

Geiger, T. and Michel, F. (2005) 'Studies on the polysaccharide JunFunori used to consolidate matt paint', *Studies in Conservation* 5(3): 193–204.

Hedlund, H.-P. and Johansson, M. (2005) 'Prototypes of Lascaux's Medium for Consolidation', *Restauro* 6: 432–9.

Schilling, M.R. and Khanjian, H.P. (1996) 'Gas chromatographic analysis of amino acids as ethyl chloroformate derivatives. Part 2. Effects of pigments and accelerated aging on the identification of proteinaceous binding media', *Journal of the American Institute for Conservation* 35(2): 123–44. Available at: http://cool.conservation-us.org/jaic/articles/jaic35-02-004.html (accessed 6 September 2012).

Author's address

Chantal Thuer (C_Thur@hotmail.com)

Coming unstuck! The approach to the treatment of a recurrently flaking Baroque ceiling painting

Zoe Roberts and John Burbidge

Introduction

The subject of this paper is one of the Baroque ceiling paintings at Hampton Court Palace (Surrey, UK). The painting (oil paint on lath and plaster) crowns William Kent's 1734–35 decorative scheme for the Queen's Staircase, the principal entrance to the Queen's Apartments (Figure 1).

In 2001, a campaign of treatment was undertaken to stabilise widespread scattered areas of flaking paint using a conventional wall paintings adhesive. Within nine months recurrent flaking was observed that continues to the present time, in spite of repeated localised treatment. Since 2005, this painted ceiling has been the focus of a holistic investigation that has looked into the causes of the flaking and possible solutions to the problem, of which the treatment trial work described in this paper is just one part.

Background

One of the biggest challenges with this painting is its location and size. It sits 12 m from ground level above a very busy visitor thoroughfare right at the heart of the palace.[1] The large rectangular cove ceiling measures 15.55 × 9.05 m. The painting is a splendid architectural *tromp l'oeil*, with fictive marble figures supporting a fictive cornice from which a coffered marble dome rises. The dome is decorated with rows of gilt-framed paterae alternating with wreathed decorative elements that radiate from a central ellipse containing the insignia of the Order of the Garter (Croft-Murray 1970: 233).

In 2001, the preventive conservation team at Hampton Court Palace, whilst working in the vicinity of the painting, noticed flaking across the surface of the ceiling (Figure 2). An inspection by Granville & Burbidge[2] revealed this flaking to be widespread (Burbidge 2001a). A full birdcage scaffold was installed and the flaking stabilised using Plextol B500 in deionised water (1:10 and 1:5 v/v), an acrylic dispersion commonly used for the treatment of wall paintings (Burbidge 2001b). The adhesive was applied through an intervention layer of lens tissue. Pre-wetting with industrial methylated spirit (IMS) and deionised water improved the penetration of the adhesive. The flakes were pushed back against the surface with cotton wool swabs dampened with deionised water. After systematically working over the entire ceiling, the areas that were fixed were checked and unstable areas were re-treated. In cases where the paint structure was thicker, a stronger adhesive mix was used, and an electric heated spatula was applied to the surface through a protective layer of polyester film. Only localised areas of lifting paint

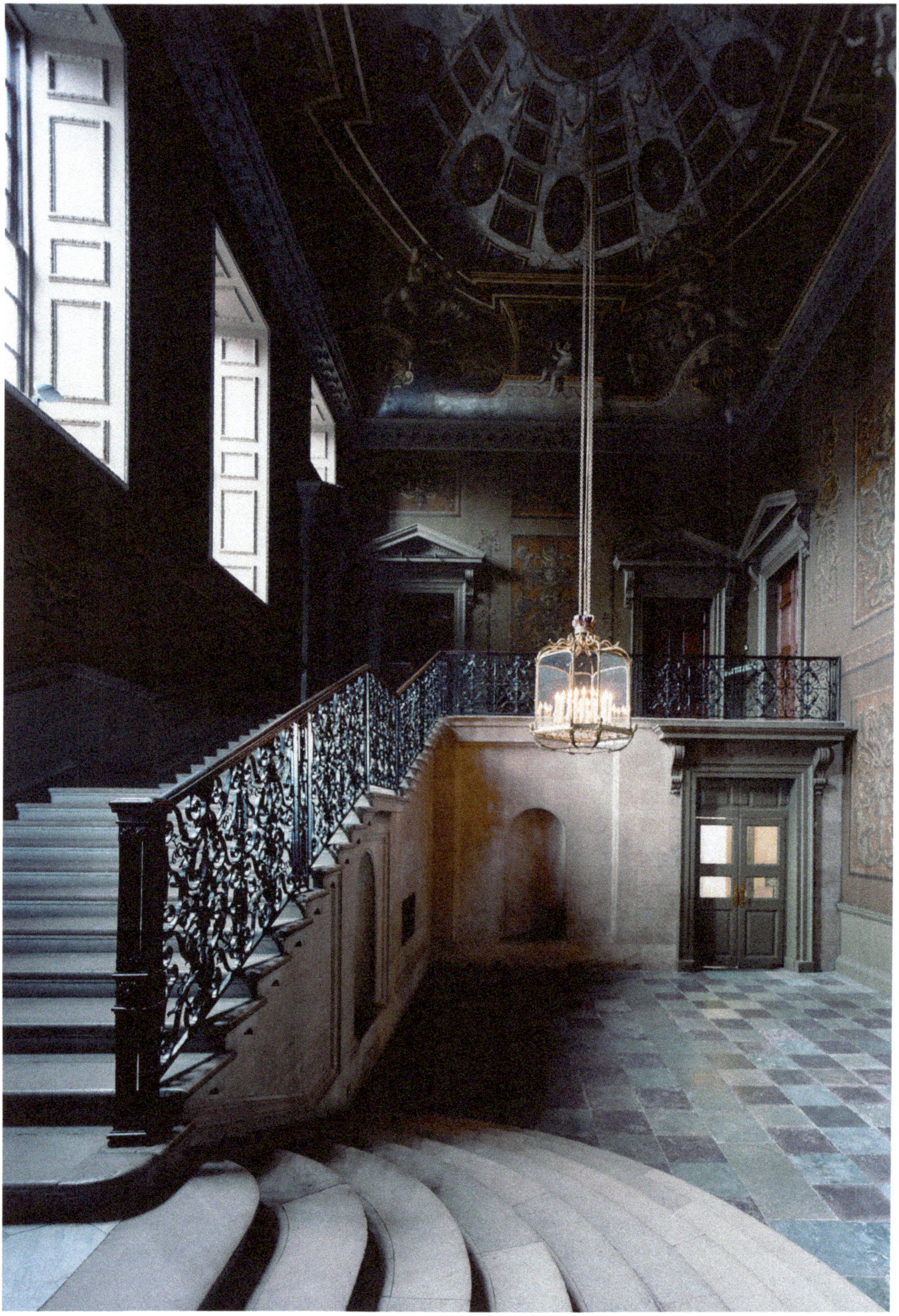

Figure 1 Queen's Stairs, Hampton Court Palace.

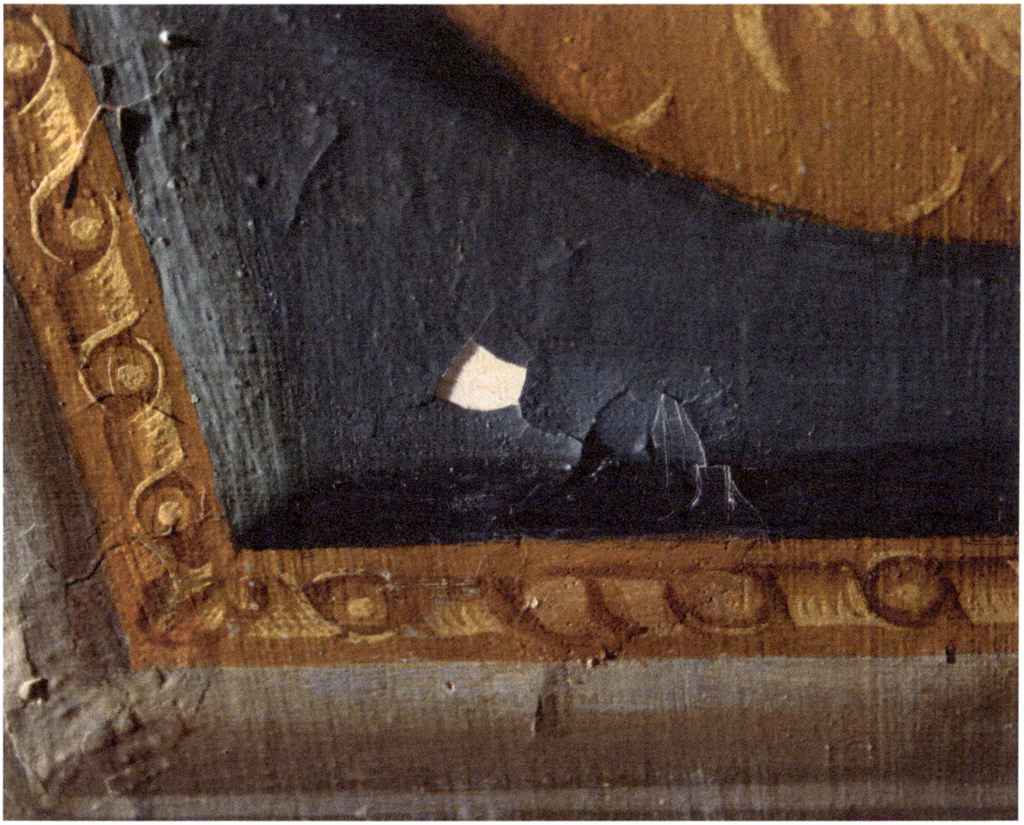

Figure 2 An example of an area of flaking paint including an area of loss.

were fixed – the ceiling was not given blanket treatment.

In addition to the physical awkwardness of stabilising paint on a ceiling, the paint was found to be brittle, slightly cupped and distorted, so the numerous lifting flakes no longer fitted their original lacunae. When pressed back fractures occurred in the paint structure, sometimes resulting in minor losses (Figure 3). Moreover, because the paint flakes often overlapped when pressed back, this reduced the area of contact available for readhesion.

Whilst undertaking an audit of the condition of the wall paintings in the staircase nine months later in 2002, there was also an opportunity to review briefly the condition of the ceiling painting from a mobile elevating work platform. Once again a number of areas of flaking paint were found on the ceiling (Burbidge 2002). In most areas the lifting paint appeared to be a recurrence rather than a new phenomenon, visible by telltale signs of fractures and displacement caused during fixing.

In 2003, there was a chance to undertake a single day of treatment to fix the recurrent flaking (Burbidge 2003). On inspection, the condition of the painting was worse than expected, with examples of flaking paint more extensive than observed during the last inspection. Widespread cleavage between the paint structure and the plaster support was visible. Some of this seemed to be new, yet the majority appeared again to be a recurrence of lifting in areas that had previously been stabilised.

By this point it was realised that the problems of the ceiling painting were more complex than first envisaged and further research would be needed to implement a solution. This

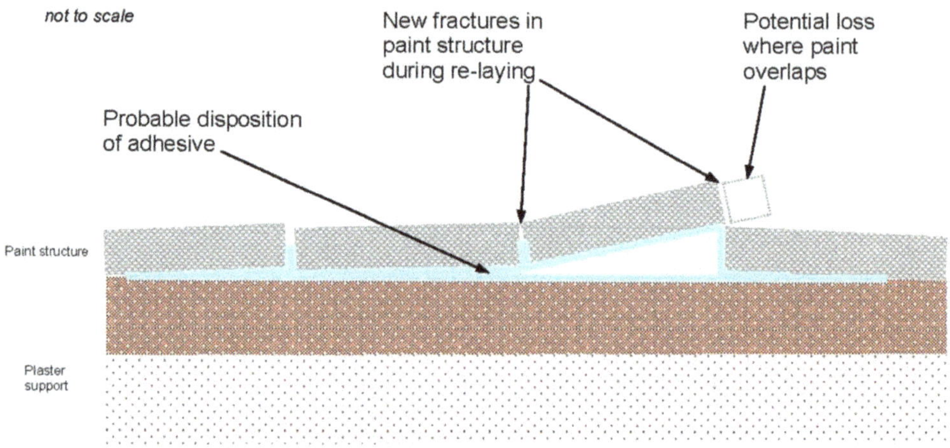

Figure 3 Schematic cross-section through lifting/flaking paint illustrating potential damage and reason for future failure as a result of fixing.

resulted in the following lines of investigation: paint analysis; archival research into the history of the painting and previous restoration campaigns; environmental monitoring and investigation and treatment trials.

Paint analysis

Paint analysis revealed an underlying, poorly bound/ unbound chalk layer at the interface with the plaster as the point of cleavage (Eastaugh 2005). This unbound chalk layer probably represents the original whitewash finish of the ceiling when the space was newly built by Sir Christopher Wren in 1694 (Davies 2005, 2006). It is thought that this layer was not brushed back 40 years later in 1734, when Kent was commissioned by George II and Queen Caroline to create a more impressive entrance (Thurley 2003). However, we have recently also found this same layer structure in Kent's Cupola room ceiling at Kensington Palace, London. Iconographically the Kensington Palace ceiling is very similar, but has in contrast proved to be stable (Davies 2011). Therefore, although this chalk layer cannot be considered the only reason for the failure we can assume that it represents an intrinsic weakness in the structure of the ceiling painting.

Archival research

Twentieth-century records and close visual examination of the painting show that the problem of the flaking is not entirely new. Our limited archival records document two major restorations since 1954. Examination of the surface revealed dislodged paint flakes, held in place by the varnish layer, indicating previous instability of the painting. The records are not specific, however, and treatments may have involved consolidation and coating with waxes, resins and varnishes alongside repainting and retouching. These restorations have added weight and complexity to the layer structure of the painting, adding to the problem of penetration for the adhesives that we currently use.

Archival records have also shown that the painting suffered from several occurrences of large-scale water infiltration through the roof of the apartment above. A particularly dramatic occurrence of this resulted in the ceiling

Table 1 Properties of the three adhesives trialled in 2004 and the results of the trial.

Product	Composition	Application	Results
Beva 371 diluted in white spirit	Hot-melt adhesive. Contains wax, ethylene/vinyl acetate, ketone resin and rosin	Applied warm, directly to the paint. Area passed over with hot spatula. Residue cleared with petroleum spirit	Good handling, enabled large flakes to be reattached
Paraloid B-72 + xylene (1:10 w/v)	Acrylic copolymer (ethyl/methyl methacrylate)		Xylene found to solubilise the overpaint – test discontinued
Plexisol P550-40 + white spirit (1:9 v/v)	Acrylic polymer (n-butylmethacrylate)	Applied directly to the paint. Area passed over with hot spatula. Residue cleared with white spirit	Solution appeared insufficiently strong enough to stabilise fragments that remained loose

Table 2 Resins considered for the second trial in 2005.

Name	Type	Molecular weight (MW)	Glass transition temperature (T_g) in °C
Beva 371	Hot-melt adhesive containing wax, ethylene/vinyl acetate, ketone resin and rosin		
Plexisol B550-40	Acrylic polymer (n-butylmethacrylate)	Low	34
Paraloid B-72	Acrylic copolymer (ethyl/methyl methacrylate	11,397	40
Regalrez 1094	Hydrogenated hydrocarbon	630	43.8
Arkon P90	Hydrogenated hydrocarbon	620	35.6
Escorez 5380	Hydrogenated polycyclopentadiene	391/349	48.6/31.5

being described as 'ill-fated' in 1971. In spite of a subsequent restoration campaign, it is thought that this has probably contributed to the instability that we see today.

In spite of misfortunes and restorations in the course of its nearly 300-year history, a very detailed condition survey in 2010, including examination in ultraviolet light and paint analysis, revealed that approximately 90% of the painted surface is original.

Environmental research

The Queen's Stairs area experiences heavily fluctuating environmental conditions due to its situation within the palace. The staircase contains two external doors at ground level and three internal ones on the first floor as well as four large windows. The lower doors are in constant use by visitor traffic as the staircase sits on a main route through the palace. Detailed environmental monitoring has shown that there are wide daily and seasonal variations in relative humidity (RH) and absolute humidity caused by high exchange rates with the external environment (Curteis 2005–2007). This erratic climate (RH yearly changes within the range 35–80% and diurnal changes of up to 15%) is likely to cause physical stress to a paint structure in good condition. Work to improve the environment within the space is ongoing.

Treatment trials

It is the policy of Historic Royal Palaces to preserve the appearance of all of its wall paintings, inclusive of superimposed layers of retouching/repainting and varnish (Frame and Julien-Lees 2007). Thus for the time being, preserving the appearance of the ceiling painting inclusive of solvent-sensitive retouching and varnishes[3] must be taken into account in the design and development of a long-term conservation strategy.

Initial selection of resins

By 2004 the failure of the Plextol B500 treatment undertaken in 2001 was acknowledged, and tests were started to find an alternative

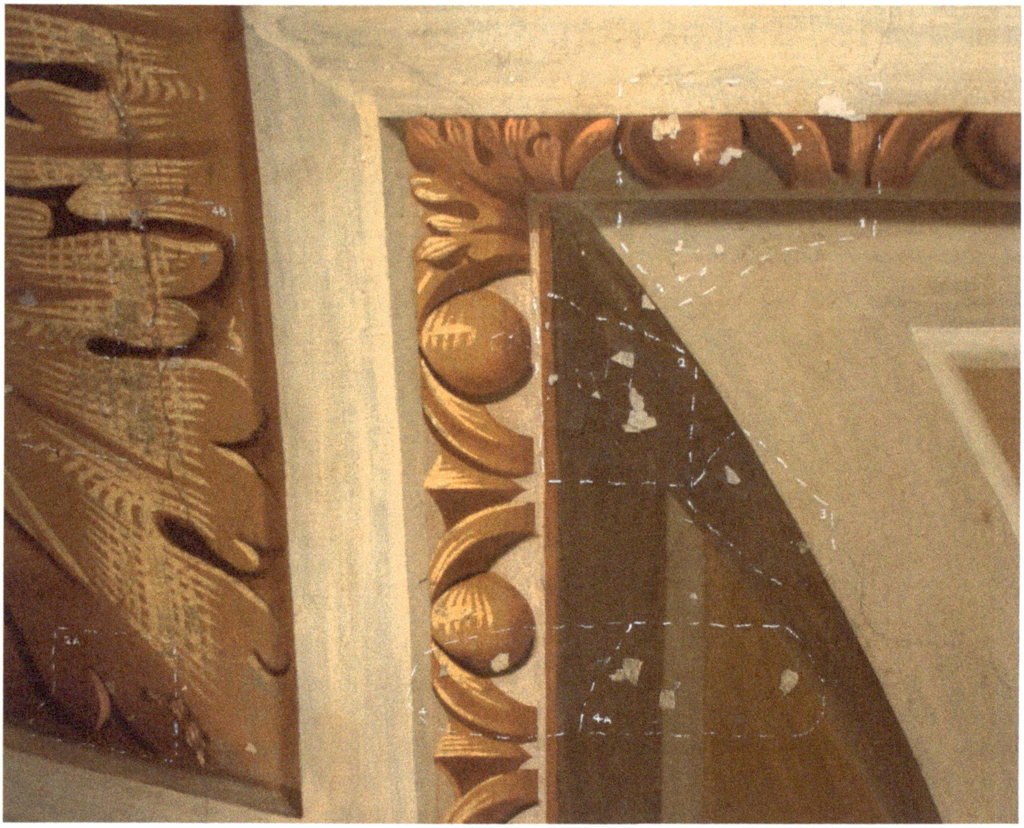

Figure 4 Detail of the coving showing the location of the fixing trials, outlined in white, after treatment.

adhesive for the lifting and flaking paint. A preliminary trial of three non-polar, solvent-based adhesives (Beva 371, Paraloid B-72, Plexisol P550-40) was made *in situ* (Burbidge 2004) (Table 1).

The trial was carried out in an inconspicuous area of retouched paint where flaking was thought to be typical of that seen over the whole ceiling. The area displayed a fairly concentrated cluster of lifting and flaking paint that would enable comparative tests to be undertaken. Of the three adhesives tested, only Beva 371 was found to be successful.

Broadening the selection of resins

By 2005, the stratigraphy of the painting was better understood and the unbound chalk layer had been identified as the point of cleavage. It was thought that previous consolidation attempts may have been unsuccessful because they had not penetrated to this chalk layer. This led to the idea of testing low molecular weight resins on the basis that these might penetrate sufficiently through the painting's complex stratigraphy due to their smaller molecular size (René de la Rie and McGlinchey 1990; René de la Rie 1993).

From an initial list of resins considered in 2004 (Beva 371, Plexisol P550-40 and Paraloid B-72) a further list of potential products was suggested by Nicholas Eastaugh (Regalrez 1094, Arkon P90 and Escorez 5380) (Burbidge 2005) (see Table 2 for their properties).

Although Paraloid B-72 had an appropriate glass transition temperature (T_g)[4] and high molecular weight, the previous trials had found

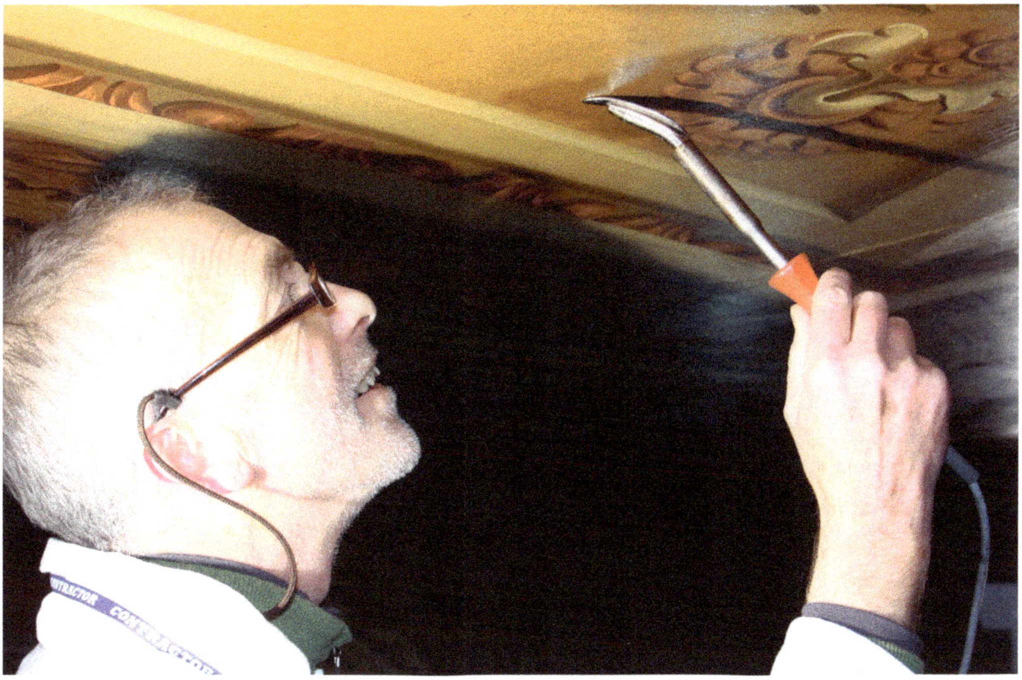

Figure 5 John Burbidge undertaking flake fixing on the Queen's Stairs ceiling.

that a solution in xylene solubilised the overpaint on the ceiling.[5] Arkon was unobtainable at the time of the trials. Escorez 5380 is used mainly as a tackifier for adhesives so it was therefore decided to use this to improve the properties of Regalrez 1094 (Whitten 1995).

Making a mock-up panel

Further tests involving Beva 371, Plexisol P500-40, Regalrez 1094, Escorez 5390 and Reglarez 1094:Escorez 5390 were carried out on a mocked-up panel. The adhesion, penetration and consolidation for each of the adhesives were assessed (Burbidge 2005).

In-situ *trials on the ceiling*

Following the trials on the mock-up panel, the most effective three resins were selected for trial on the ceiling: Beva 371, Plexisol P500-40 and Regalrez1094:Escorez 5390. In addition, a trial was undertaken using Regalrez:Escorez as a pre-consolidant followed by Beva 371. It was hoped that this permutation would combine the good adhesive properties of Beva 371 with the high penetration of Regalrez 1094:Escorez 5390. The paint treated by the Regalrez 1094:Escorez 5390 mixture alone continued to lift; its use was therefore discontinued (Figure 4).

The following year (2006) the *in situ* trials were assessed visually (Burbidge 2006). Although all trial areas demonstrated some instability (in the form of some lifting flakes), Beva 371 and the application of Regalrez 1094: Escorez 5390 followed by Beva 371 were obviously the most successful. A second trial of both was undertaken over a larger area, this time increasing the strength to 1:4 (both the Beva 371 and the Regalrez 1094: Escorez 5390) and applying the resin over the entire surface. When this was reassessed in 2007, a small area of instability was observed where

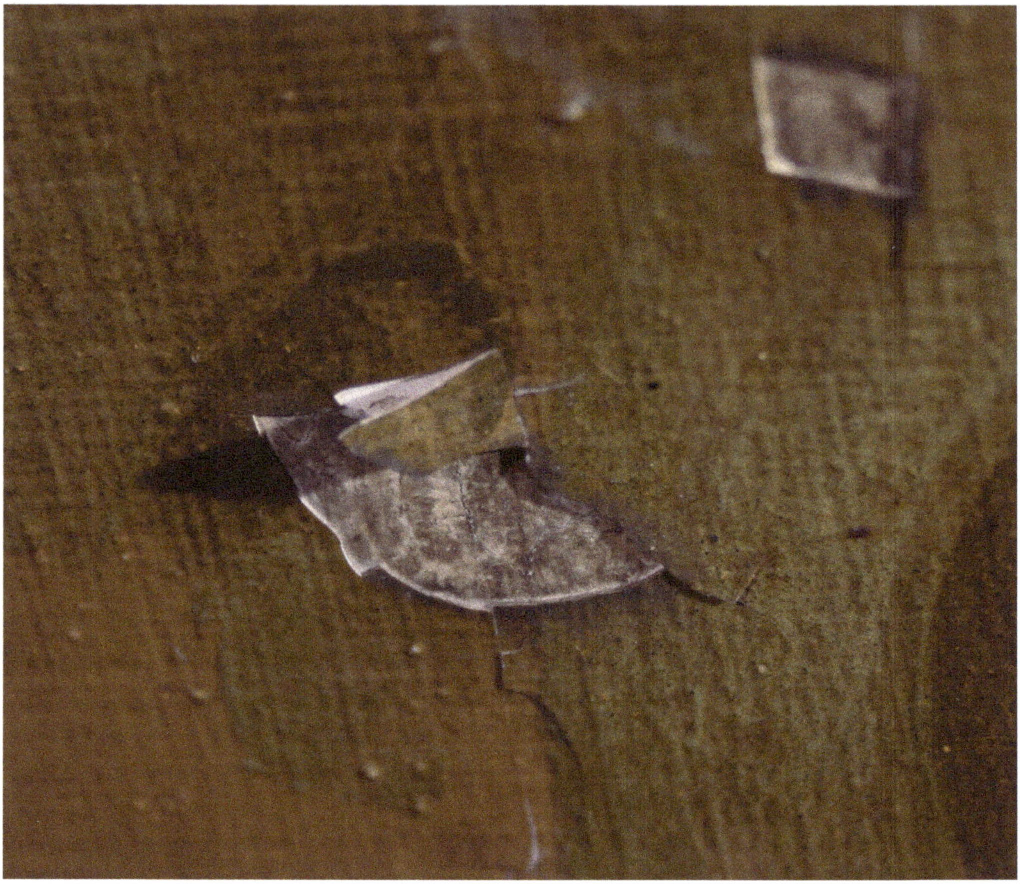

Figure 6 An area of previously re-toned loss with a new area of lifting paint pre-fixing.

Regalrez 1094:Escorez 5390 followed by Beva 371 was applied (Burbidge 2007). Subsequent monitoring of these areas has revealed a slow but persistent increase of recurrent flaking (Burbidge 2009). Areas treated with Beva 371 alone were found to be the least affected by recurrent lifting.

Treatment solution based on the result of the trials

As a result of the trial work, since 2005 emergency stabilisation of the flaking paint in imminent danger of loss has been carried out using Beva 371 alone. The adhesive is applied warm by brush directly to the flakes, which are pressed back, initially with the brush and then with a heated spatula (Figure 5). Surface residue is then cleaned with petroleum spirit. In addition, the small areas of exposed plaster caused by paint losses in the immediate vicinity are toned (Figure 6). This partial integration is undertaken so that new losses are more readily apparent. This treatment is carried out every one to two years. Because the ceiling is 12 m from ground level and because total access requires a full birdcage scaffold and incurs large-scale cost and disruption to visitors, this stabilisation is most often carried out from a mobile platform. The treatment has a good success rate with fewer than 5% of treated areas being observed to fail in subsequent assessments.

In the absence of a long-term treatment solution, we have chosen to continue with the programme of emergency stabilisation, treating flakes in imminent danger of loss. It is considered that a wholesale treatment with Beva 371 risks making the paint/ varnish layers heavier and more coherent without necessarily improving adhesion to the substrate and could therefore lead to larger areas of loss in the long term. This would also limit the application of alternative adhesives in the future.

In order to better understand, track and quantify the rate and nature of the deterioration, full rectified photography of the ceiling was commissioned in 2010 and a detailed condition survey was undertaken (Burbidge 2010).

Conclusions

As the title of this paper suggests, a permanent treatment solution to the problem of the flaking paint has not been found as yet; the current condition of the ceiling is reminiscent of its state in 2001. Since 2004, the deterioration mechanisms have been investigated, a variety of materials has been tested as consolidants and different adhesive application techniques have been considered. Even though Beva 371 generally appears to be successful when locally applied, ongoing treatment with this adhesive is not necessarily a long-term answer to the problem.

It is clear that the deterioration of the painted ceiling has very complex causes. The root of the instability is a combination of intrinsic weakness from manufacture, damage from water ingress, as well as modification following restoration treatments. Moreover, the problems of the painting are clearly exacerbated by the unstable microclimate within the staircase. As this is something that we may be able to change, over the past three years we have concentrated on gaining more control of this environment with the intention of slowing the rate of deterioration and prolonging the life of the Beva 371 treatment.

As the painting is located within a Scheduled Ancient Monument and forms part of a significant architectural space, studies needed to be undertaken before any major changes either to historic fabric or to visitor routing would be supported by internal and external stakeholders. A variety of solutions has been considered, including glazing the ceiling, visitor rerouting, a double-door system and air curtains above the external doors. A programme of testing has also been carried out to establish if it would be practically possible to reduce air exchange and therefore stabilise the internal microclimate whilst minimising intervention to the historic fabric and still facilitating visitor access. This involved both pressure testing the space as well as using environmental modelling (Roberts *et al.* 2010).

As a result of this work, in the last year all apertures have been draught-proofed and a draught lobby has been added on the outside of the set of external doors that were shown to have the greatest influence on destabilising the internal environment. Environmental data that will demonstrate the result of this intervention are currently being collected. The environmental work, along with regular monitoring of the condition and emergency treatment, is very much ongoing.

Materials and suppliers

- Plextol B500 (aqueous dispersion of a copolymer based on ethyl acrylate and methyl methacrylate): Lascaux Colours & Restauro, Barbara Diethelm AG, Zürichstrasse 42, 8306 Brüttisellen, Switzerland
- Beva 371 (hot-melt adhesive containing wax, ethylene/vinyl acetate, ketone resin and rosin): Archival Aids/Particle Technology Limited, Station Yard Industrial Estate, Hatton, Derby DE65 5DU, UK
- Plexisol B550-40 (acrylic polymer (n-butylmethacrylate)
- Paraloid B-72: Rohm and Haas

- Regalrez 1094 (hydrogenated hydrocarbon): Eastman Chemical Company, 200 South Wilcox Drive, Kingsport, TN 37660-5280, USA
- Arkon P-90 (hydrogenated hydrocarbon): Arakawa Europe GmbH, Am Kronberger Hang 4, D-65824 Schwalbach/Ts (unavailable for the trial)
- Escorez 5380 (hydrogenated polycyclopentadiene): ExxonMobile Chemical Company, 13501 Katy Freeway, Houston, TX 77079-1398, USA
- PEL Wet Strength Repair Tissue: Preservation Equipment Limited, Vinces Road, Diss, Norfolk IP22 4HQ, UK

Acknowledgements

We would like to thank Kate Frame and colleagues in Conservation and Collection Care, Historic Royal Palaces especially Kathryn Hallett and Sophie Julien-Lees. In particular, we would also like to thank external consultants: Tobit Curteis, Jane Davies, Dr Nick Eastaugh and Dr Spike Bucklow.

Notes

1. Hampton Court Palace is open 364 days a year and visitors generate approximately 60% of Historic Royal Palaces income.
2. Granville & Burbidge Conservation and Restoration is a small private company specialising in wall paintings and polychrome sculpture.
3. The painting's varnish is readily soluble in acetone and alcohol.
4. The glass transition temperature (T_g) is the temperature at which a polymer changes from a glassy state to a rubbery one and determines many of the properties of the material (Horie 1998: 18). Polymers with a T_g in the region of 40°C were considered to have the appropriate amount of flexibility for the situation of this painting.
5. An *in situ* trial in 2010 substituting the solvent 1-methoxy-2-propanol gave similar results.

References

Burbidge, J. (2001a) *Emergency Inspection of the Ceiling (31.01.01), Queen's Staircase, Hampton Court Palace*. Granville & Burbidge, 19 February 2001. Unpublished report.

Burbidge, J. (2001b) *Treatment of the Painted Ceiling (16–25.07.01), Queen's Staircase, Hampton Court Palace*. Granville & Burbidge, 15 August 2001. Unpublished report.

Burbidge, J. (2002) *Condition Audit of the Painted Walls (15.04.02), Queen's Staircase, Hampton Court Palace*. Granville & Burbidge, 28 June 2002. Unpublished report.

Burbidge, J. (2003) *Treatment of the Painted Ceiling (14.01.03), Queen's Staircase, Hampton Court Palace*. Granville & Burbidge, 25 March 2003. Unpublished report.

Burbidge, J. (2004) *Condition Assessment of the Painted Ceiling (24/25.02.04), Queen's Staircase, Hampton Court Palace*. Granville & Burbidge, 5 May 2004. Unpublished report.

Burbidge, J. (2005) *Painted Ceiling: Emergency Treatment and Resin Trials (1–3 June 2005), Queen's Staircase, Hampton Court Palace*. Granville & Burbidge, 26 July 2005. Unpublished report.

Burbidge, J. (2006) *Painted Ceiling: Assessment of Resin Trials and Emergency Treatment 20/21st February 2006, Queen's Staircase, Hampton Court Palace'*. Granville & Burbidge, 7 April 2006. Unpublished report.

Burbidge, J. (2007) *Painted Ceiling: Emergency Treatment 21st & 22nd March 2007, Queen's Staircase, Hampton Court Palace*. Granville & Burbidge, 28 May 2007. Unpublished report.

Burbidge, J. (2009) *Painted Ceiling: Emergency Treatment 18th–20th March 2008, Queen's Staircase, Hampton Court Palace*. Granville & Burbidge, 17 September 2009. Unpublished report.

Burbidge, J. (2010) *Painted Ceiling Survey 13th January–6th February 2010. Volume 1, Queen's Staircase, Hampton Court Palace*. Granville & Burbidge, 31 May 2010. Unpublished report.

Croft-Murray, E. (1970) *Decorative Painting in England 1537–1837*, vol. 2. Feltham: Country Life Books.

Curteis, T. (2005–2007) *Environmental Survey of the Ceiling Painting Years 1–3*. Tobit Curteis Associates. Unpublished reports.

Davies, J. (2005) *Draft Report on Paint Sample Examination, Queen's Staircase, Hampton Court Palace January 2005*. Jane Davies Conservation. Unpublished report.

Davies, J. (2006) 'Microscopy and archival research: interpreting results within the context of historical records and traditional practice', *infocus* 3: 32–45.

Davies, J. (2011) *William Kent Ceiling Paintings at Hampton Court Palace and Kensington Palace: Architectural Paint Research 2010*. Jane Davies Conservation. Unpublished report.

Eastaugh, N. (2005) *Scientific Examination Report: The William Kent Ceiling, Queen's Staircase, Hampton Court Palace. 25th January 2005*. Unpublished report.

Frame, K. and Julien-Lees, S. (2007) 'Mural paintings within the Historic Royal Palaces: our approach to their continuous care', in R. Gowing and R. Pender (eds), *All Manner of Murals: The History, Techniques and Conservation of Secular Wall Paintings*. London: Archetype Publications, 69–78.

Horie, C.V. (1998) *Materials for Conservation: Organic Consolidants, Adhesives and Coatings*. Oxford: Architectural Press.

René de la Rie, E. (1993) 'Polymer additives for synthetic low-molecular-weight varnishes', in J. Bridgland (ed.), *Preprints of ICOM-CC 10th Triennial Meeting*. London: James & James, 566–73.

René de la Rie, E. and McGlinchey, C.W. (1990) 'New synthetic resins for picture varnishes', in *Cleaning, Retouching and Coatings: Technology and Practice for Easel Paintings and Polychrome Sculpture. Preprints of the Contributions to the Brussels Congress, 3–7 September 1990*. London: International Institute for Conservation of Historic and Artistic Works, 168–73.

Roberts, Z., Curteis, T. and Gibb, I. (2010) 'Balancing environmental needs with visitor access: a case study of the Queen's Stairs at Hampton Court Palace', paper presented at ICONs CF10 *Conservation in Focus*, 24 March 2010.

Thurley, S. (2003) *Hampton Court: A Social and Architectural History*. New Haven, CT and London: Yale University Press, 277–9.

Whitten, J. (1995) 'Regalrez 1094: properties and uses', *WAAC Newsletter* 17(1): 11–12.

Author's address

Zoe Roberts, Conservation and Collection Care, Historic Royal Palaces, Hampton Court Palace, East Molesey, Surrey KT8 9AU, UK (zoe.roberts@hrp.org.uk)

Development of a gluing jig for use with small panels

Fabio Mazzochini

The use of purpose-built jigs for joining and gluing panels has been developed for use in museums over the past 50–60 years. However the size and cost of the jigs have restricted their use in small institutions and private studios. This paper discusses the development of a compact but multifunctional jig, designed for use in a small workshop space.

A large proportion of the panel paintings treated in private studios such as mine are usually of relatively small dimensions – no more than 120 cm in either direction. The treatment of panels often necessitates the removal of restrictive cradles and the manipulation of joints to reduce distortions. Simple wood clamps and braces are often used to achieve the best alignment of joints during re-gluing, however positioning of these instruments on a work surface can often be complicated and may obstruct a good view of the surfaces requiring realignment.

Conservation departments in museums such as the National Gallery in London have developed gluing jigs to facilitate structural work on panels (Reeve 1990; Kozlowski 1962). This type of jig is usually large so it can accommodate the size of panels held in museum collections and is likely to take up a large permanent footprint in the studio. Therefore, the challenge presented to me was to make a jig that incorporated elements of the large gluing jigs, but to adapt it to function in the restricted space of a small studio. The criteria set for the new jig were:

- it should be relatively lightweight and easily movable
- it should allow access to both sides of a panel
- that once clamped in the jig the panel can be moved through a range of angles
- it should be constructed relatively cheaply and easily.

The jig was made from aluminium as it is light, strong and relatively easy to work with. All the materials used to make the jig were readily available from hardware stores and ironmongers with the larger sections of aluminium sourced from an aluminium factory. No welding was used in the jig's construction. All the joints were assembled using nuts, bolts and rivets. A router was used to make grooves in the aluminium section and a few hand tools were used to work the aluminium.

The jig stands like an easel and measures 180 × 130 cm when in vertical orientation and 120 × 130 cm when in horizontal orientation (Figure 1). It weighs approximately 25 kg. The clamped section can be tilted through 180 degrees and stabilised in four different positions (Figure 2) facilitating access to manipulate alignment (Figure 3) and the use of raking light to check the alignment of the joint (Figure 4). The two sections of a panel can be clamped separately and then brought together. This allows the alignment of the joint to be checked before gluing. Once the correct position is established the edges can be opened

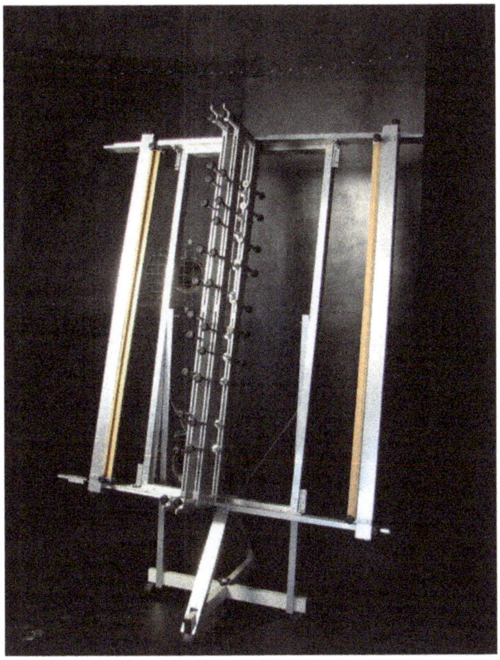

Figure 1 The jig.

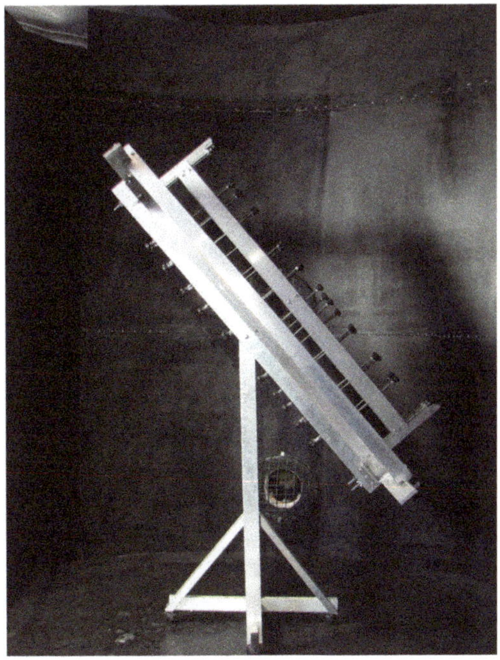

Figure 2 The jig at an angle during rotation.

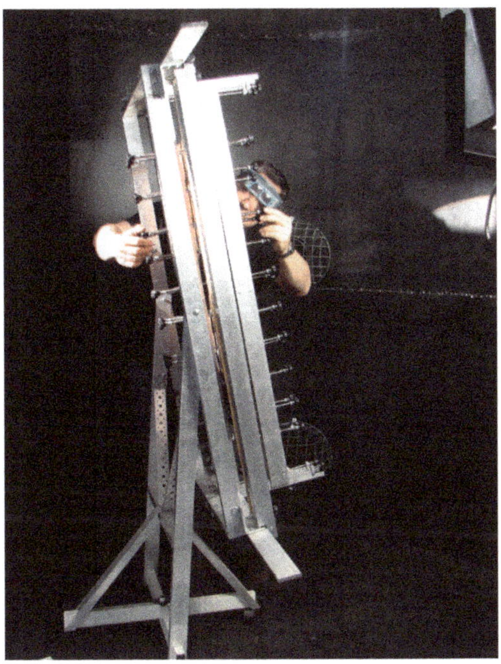

Figure 3 Moving the clamps to adjust alignment.

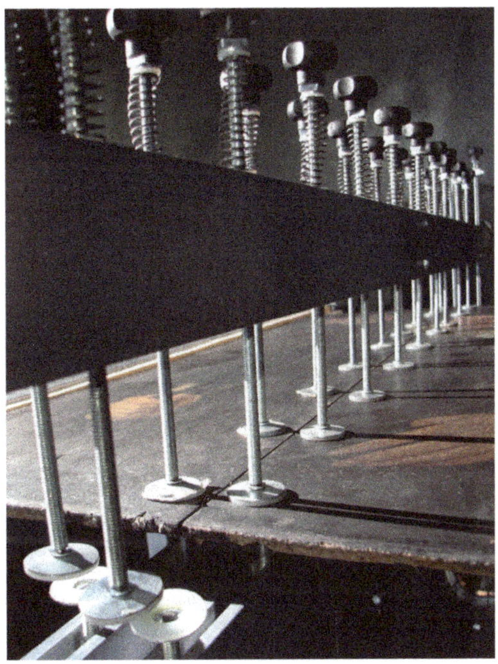

Figure 4 Panel in the jig under raking light.

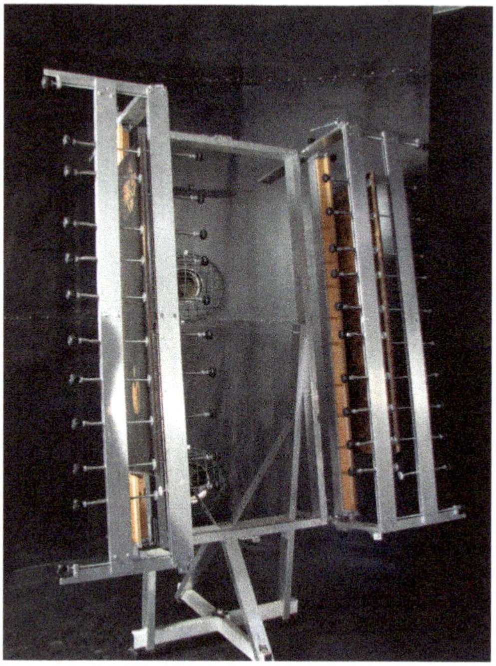

Figure 5 The jig in an open position.

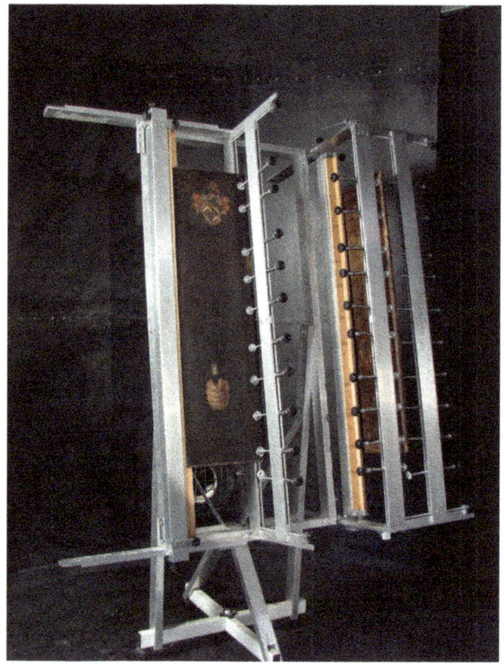

Figure 6 The jig with two parts of panel held in different alignments.

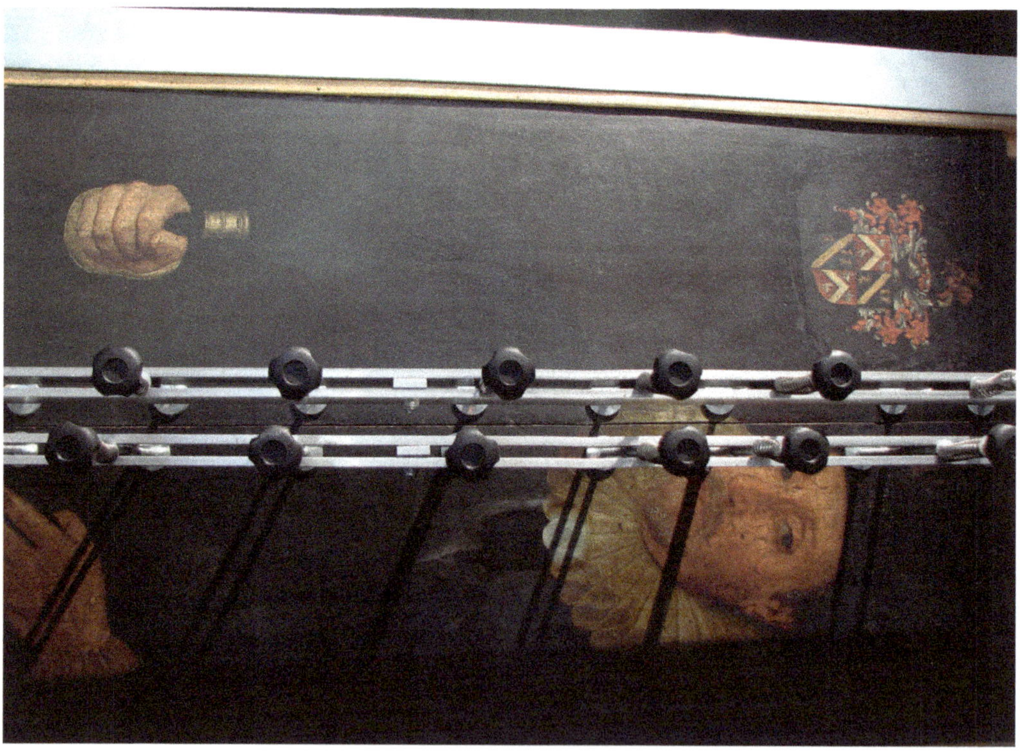

Figure 7 The panel in the jig in the process of aligning edges.

to allow application of the glue then brought back to exactly the same position and clamped in place (Figures 5–7).

The machine has four main mechanical actions:

1. It tilts in order to use raking light to observe the right alignment of the front in four different positions, including with the panel upright, as it would be under normal display conditions.
2. It aligns the panel using the sliding clamps disposed along the clamping bars.
3. It opens the two sections of the panel after alignment in order to apply the adhesive along the joining edges of the panel.
4. It clamps the panel and presses together the two edges to allow curing of the glue for at least 24 hours.

In order to support the outer edges of the planks and ensure the right curving of the panel alongside the two clamping bars, a plank of flexible plywood L-shaped section is inserted where the edges of the two panel sections are sitting. These two plywood parts can slide forwards and backwards, controlled by four knobs placed at the extremity of the clamping bars.

Although taking elements from current designs of other panel rejoining jigs, this new design has provided me with more flexibility and manoeuvrability within a private studio setting, whilst maintaining a high level of functionality.

References

Kozlowski, R. (1962) 'An apparatus for gluing split panels', *Studies in Conservation* 7(4): 135–40.

Reeve, A.M. (1990) 'A new multi-purpose clamping table for the treatment of paintings on wood', *Studies in Conservation* 35: 160–62.

Author's address

Fabio Mazzocchini, Unit 22–23 Sulivan Enterprise Centre, Sulivan Road, London SW6 3DJ, UK (mazzocchinibean@aol.com)

www.ingramcontent.com/pod-product-compliance
Ingram Content Group UK Ltd.
Pitfield, Milton Keynes, MK11 3LW, UK
UKHW061056180826
474418UK00006B/56

9 781904 982883